Test Item File

Calculus 6e

Edwards & Penney

Prentice Hall

Upper Saddle River, NJ 07458

Acquisitions Editor: Eric Frank
Supplement Editor: Aja Shevelew
Assistant Managing Editor: John Matthews
Production Editor: Wendy A. Perez
Supplement Cover Manager: Paul Gourhan
Supplement Cover Designer: Joanne Alexandris
Manufacturing Buyer: Ilene Kahn

 © 2002 by Pearson Education, Inc.
Pearson Education, Inc.
Upper Saddle River, NJ 07458

All rights reserved. No part of this book may be reproduced in any form or by any means, without permission in writing from the publisher.

The author and publisher of this book have used their best efforts in preparing this book. These efforts include the development, research, and testing of the theories and programs to determine their effectiveness. The author and publisher make no warranty of any kind, expressed or implied, with regard to these programs or the documentation contained in this book. The author and publisher shall not be liable in any event for incidental or consequential damages in connection with, or arising out of, the furnishing, performance, or use of these programs.

Printed in the United States of America

10 9 8 7 6 5 4 3 2

ISBN 0-13-066162-7

Pearson Education Ltd., *London*
Pearson Education Australia Pty. Ltd., *Sydney*
Pearson Education Singapore, Pte. Ltd.
Pearson Education North Asia Ltd., *Hong Kong*
Pearson Education Canada, Inc., *Toronto*
Pearson Educacíon de Mexico, S.A. de C.V.
Pearson Education—Japan, *Tokyo*
Pearson Education Malaysia, Pte. Ltd.
Pearson Education, *Upper Saddle River, New Jersey*

CONTENTS

Chapter	1	Functions, Graphs, and Models	1
Chapter	2	Graphs of Equations and Functions	30
Chapter	3	The Derivative	42
Chapter	4	Additional Applications of the Derivative	69
Chapter	5	The Integral	112
Chapter	6	Applications of the Integral	144
Chapter	7	Calculus of Transcendental Functions	160
Chapter	8	Techniques of Integration	174
Chapter	9	Differential Equations	191
Chapter	10	Polar Coordinates and Parametric Curves	197
Chapter	11	Infinite Series	240
Chapter	12	Vectors, Curves, and Surfaces in Space	262
Chapter	13	Partial Differentiation	272
Chapter	14	Multiple Integrals	283
Chapter	15	Vector Calculus	295

Ch. 1 Functions, Graphs, and Models
1.1 Functions and Mathematical Modeling
Solve the problem.

1) Given $f(x) = (x - 1)^3$, find: (a) $f(3)$; (b) $f(a + 1)$; (c) $f(x^2)$; (d) $f(2a)$.

   Answer: (a) 8; (b) a^3; (c) $(x^2 - 1)^3$; (d) $(2a - 1)^3$

2) Given $g(x) = x^3 - 1$, find: (a) $g(-1)$; (b) $g(3a)$; (c) $g(x + h) - g(x)$.

   Answer: (a) -2; (b) $27a^3 - 1$; (c) $(x + h)^3 - x^3$

3) Find all values of a such that $g(a) = 4$:

   (a) $g(x) = \dfrac{1}{3x - 2}$; (b) $g(x) = \sqrt[3]{x + 36}$; (c) $g(x) = 3x^2 - x + 2$.

   Answer: (a) $a = \dfrac{3}{4}$; (b) $a = 28$; (c) $a = -\dfrac{2}{3}, 1$

4) Given $f(x) = x^2 + 3x$, compute and simplify the quantity $f(a + h) - f(a)$.

   Answer: $h^2 + 2ah + 3h$

5) Find the range of values given the function $f(x) = \begin{cases} \dfrac{|x|}{x} & \text{if } x \neq 0; \\ 0 & \text{if } x = 0 \end{cases}$

   Answer: $-1, 0,$ and 1

6) The function $f(x)$ is the cost of a long distance phone call x minutes long, $0 \leq x \leq 60$. For ABC Long Distance, the cost for such a call was a 50 cent connection charge plus 7 cents for each minute the phone call lasts. Find the range of values for this function.

   Answer: [$0.50, $4.20]

7) Find the largest domain (of real numbers) on which the given formula determines a (real-valued) function:

   (a) $f(x) = x^3 - 3x$; (b) $g(x) = \dfrac{1}{\sqrt{x^2 + 1}}$; (c) $h(t) = \sqrt{t} + t^{-2}$; (d) $p(z) = \sqrt[3]{z}$

   Answer: (a) $\mathcal{R}$ (b) $\mathcal{R}$ (c) $(0, \infty)$ (d) $\mathcal{R}$

8) Find the largest domain (of real numbers) on which the given formula determines a (real-valued) function:

   (a) $f(x) = \dfrac{5}{4 - x}$; (b) $g(x) = \sqrt{9 - \sqrt{x}}$; (c) $h(t) = \sqrt{\dfrac{x - 2}{x + 2}}$

   Answer: (a) $(-\infty, 4) \cup (4, \infty)$; (b) $[0, 81]$; (c) $(-\infty, -2) \cup [2, \infty)$

9) A farmer uses 1000 meters of fencing to build a rectangular corral. One side of the corral has length x. Express the area A of the corral as a function of x.

   Answer: $A = 500x - x^2$ $(0 < x < 500)$

Calculus 1

10) A triangle has sides of lengths 3, 4, and 5. A rectangle is inscribed in the triangle with the base of the rectangle on the long side of the triangle and a corner on each of the other two sides. Express the area A of the rectangle as a function of its height x.

Answer: $A(x) = 5x - \frac{25}{12}x^2,\ 0 < x < \frac{12}{5}$

11) A regular hexagon is inscribed in a circle of radius r. Express the area A of the hexagon as a function of r.

Answer: $A = \frac{3\sqrt{3}\,r^2}{2},\ r > 0$

12) Express the area A of an equilateral triangle as a function of the length s of each of its sides.

Answer: $A = \frac{\sqrt{3}\,s^2}{4},\ s > 0$

13) Express the distance between the point (3, 0) and the point $P(x, y)$ of the parabola $y = x^2$ as a function of x.

Answer: $D = \sqrt{x^4 + x^2 - 6x + 9}$

14) The sum of the radii of two circles is 10. Express the total area A of the two circles as a function of the radius r of one of the circles.

Answer: $A = \pi r^2 + \pi(10 - r)^2,\ 0 < r < 10$

15) Use the method of repeated tabulation to approximate the solution to the quadratic equation $x^2 + 4x + 2 = 0$ that is in the interval [0, 1]. Round the answer to two decimal places.

Answer: $x = 0.59$

16) Use the method of repeated tabulation to approximate the solution to the quadratic equation $2x^2 - 8x + 5 = 0$ that is in the interval [3, 4]. Round the answer to two decimal places.

Answer: $x = 3.22$

17) Use the method of repeated tabulation to approximate the solution to the quadratic equation $3x^2 + 17x - 31 = 0$ that is in the interval [−8, −7]. Round the answer to two decimal places.

Answer: $x = -7.12$

1.2 Graphs of Equations and Functions
Solve the problem.

1) Write the point-slope equation of the straight line passing through (2, 3) and (4, 7).

Answer: $y - 3 = 2(x - 2)$

2) Write an equation of the vertical line through (3, −5).

Answer: $x = 3$

3) Write the point-slope equation of the straight line with slope −2 and y-intercept (0, 5).

Answer: $y - 5 = -2(x - 0)$

4) Write the slope-intercept equation of the straight line through the origin (0, 0) and perpendicular to the line with equation $x + 3y = 5$.

Answer: $y = 3x$

5) Write an equation of the horizontal line that passes through (-6, -2).

   Answer: $y = -2$

6) Write an equation of the line that passes has x-intercept -4 and y-intercept 5.

   Answer: $y = \dfrac{5}{4}x + 5$

7) Write an equation of the line that passes passes through (3, 2) and is parallel to the line with equation $3x + y = 8$.

   Answer: $y = -3x + 11$

8) Sketch the translated circle $x^2 + y^2 = 10x$. Indicate the center and radius.

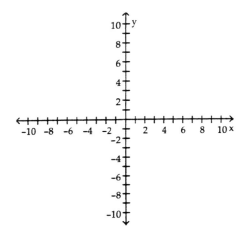

Answer:

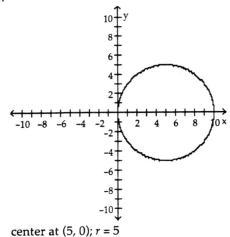

center at (5, 0); $r = 5$

9) Sketch the translated circle $x^2 + y^2 + 4x - 14y + 49 = 0$. Indicate the center and radius.

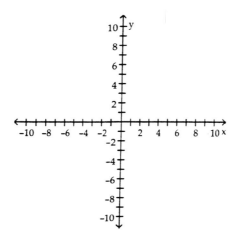

Answer:

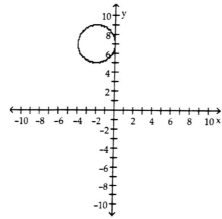

center at $(-2, 7)$; $r = 2$

10) Sketch the translated circle $4x^2 + 4y^2 - 4x + 12y = 15$. Indicate the center and radius.

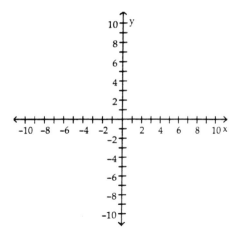

Answer:

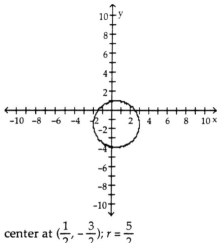

center at $(\frac{1}{2}, -\frac{3}{2})$; $r = \frac{5}{2}$

11) Write an equation of the circle with center (2, −3) and radius 6.
    Answer: $(x - 2)^2 + (y + 3)^2 = 36$

12) Find the center and radius of the circle with equation $x^2 - 6x + y^2 + 4y = 3$.
    Answer: center: (3, −2), $r = 4$

13) Find the center and radius of the circle with equation $x^2 + 4x + y^2 - 6y + 12 = 0$.
    Answer: center (−2, 3), $r = 1$

14) Sketch the translated parabola $f(x) = x^2 - 2x + 2$. Indicate the vertex of the parabola.

Answer:

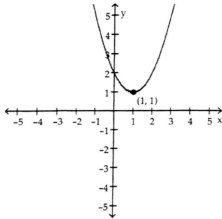

15) Sketch the translated parabola $y = 3x^2 + 12x + 14$. Indicate the vertex.

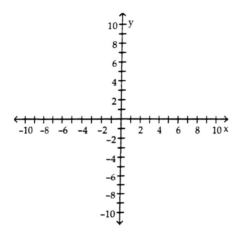

Answer:

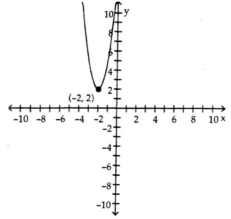

vertex at (−2, 2)

16) Sketch the graph of the function $f(x) = 5 - 2x$, $-2 < x \le 4$. Take into account the domain of the function.

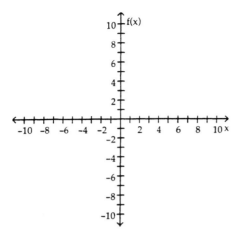

Answer:

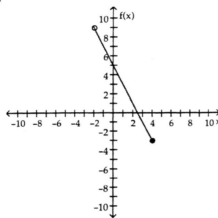

17) Sketch the graph of the function $f(x) = 8 - x^2$. Take into account the domain of the function.

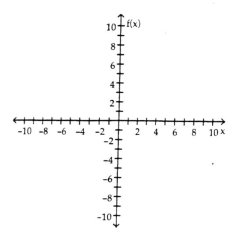

Answer:

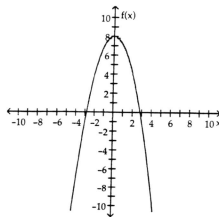

18) Sketch the graph of the function $f(x) = -x^3$. Take into account the domain of the function.

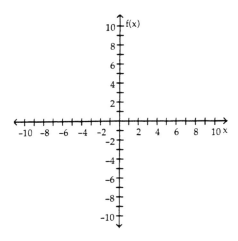

Answer:

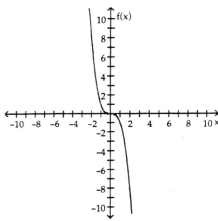

19) Sketch the graph of the function $f(x) = \sqrt{36 - x^2}$. Take into account the domain of the function.

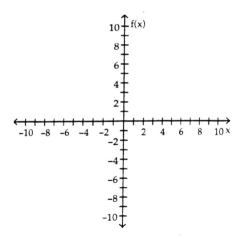

Answer:

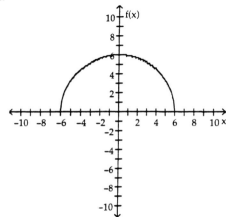

20) Sketch the graph of the function $f(x) = \dfrac{1}{x+5}$. Take into account the domain of the function.

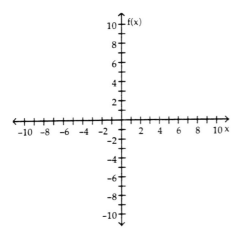

Answer:

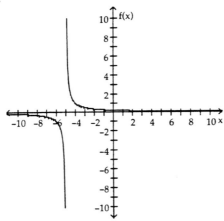

21) Sketch the graph of $f(x) = 2x + |x|$, with special attention to its shape at and near $(0, 0)$. Take into account the domain of the function.

Answer:

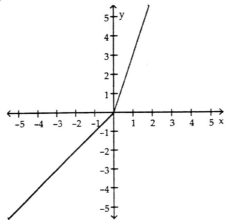

22) Sketch the graph of the function $f(x) = y = \dfrac{|x|}{x}$. Indicate any points of discontinuity.

Answer:

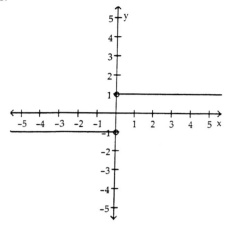

discontinuity at $x = 0$

23) Sketch the graph of the function $f(x) = \begin{cases} 1 & \text{if } x < 0 \\ -1 & \text{if } x \geq 0 \end{cases}$. Indicate any points of discontinuity.

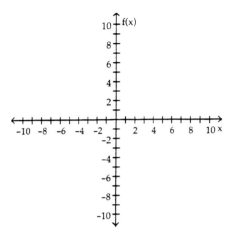

Answer:

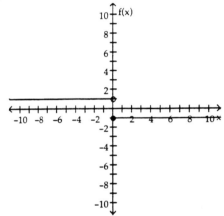

discontinuity at $x = 0$

24) Sketch the graph of the function $f(x) = [\![x]\!] + x$. Indicate any points of discontinuity.

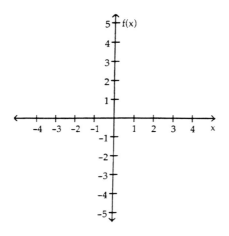

Answer:

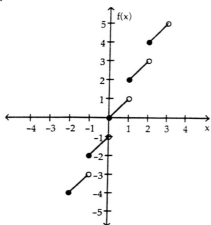

discontinuity at $x = n$, where n is an integer

25) Use a graphing utility to find (by zooming) the lowest point on the parabola $y = 2x^2 - 11x + 3$. Determine the coordinates of this point correct to two decimal places.

Answer: (2.75, −12.12)

26) Use a graphing utility to find (by zooming) the highest point on the parabola $y = 64x - 4x^2$. Determine the coordinates of this point correct to two decimal places.

Answer: (8, 256)

27) Use a graphing utility to find (by zooming) the highest point on the parabola $y = x - 3x^2$. Determine the coordinates of this point correct to two decimal places.

Answer: (0.17, 0.08)

28) A farmer has a long straight wall and 400 meters of fencing. What is the maximum rectangular area she can enclose using part of the wall to form one side of the rectangle?

Answer: 20,000 m^2

29) A rectangle in the first quadrant has one side on the x-axis, one on the y-axis, and the corner opposite the origin on the graph of the straight line with equation $y = 10 - x$. What is the maximum possible area of such a rectangle?

Answer: 25

30) Write a symbolic description of the function whose graph is pictured below.

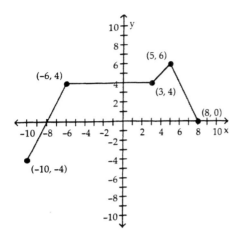

Answer: $y = \begin{cases} 2x + 16 & \text{if } -10 \leq x < -6 \\ 4 & \text{if } -6 \leq x < 3 \\ x + 1 & \text{if } 3 \leq x < 5 \\ -2x + 16 & \text{if } 5 \leq x \leq 8 \end{cases}$

31) You are traveling along a straight road connecting your home and a friend's home 150 miles apart. You drive for half an hour at 60 mph, then suddenly remember that you forgot to feed your dog. You drive back home at 60 mph. You spend half an hour at home feeding the dog, cleaning up after her, and letting her outside. You then get back in your car and drive for two and a half hours toward your original destination. Sketch the graph of the distance x (in miles) from your home as a function of time elapsed t (in hours). Also describe the function $x(t)$ symbolically.

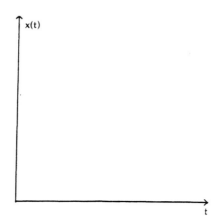

Answer:

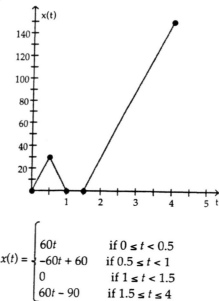

$$x(t) = \begin{cases} 60t & \text{if } 0 \le t < 0.5 \\ -60t + 60 & \text{if } 0.5 \le t < 1 \\ 0 & \text{if } 1 \le t < 1.5 \\ 60t - 90 & \text{if } 1.5 \le t \le 4 \end{cases}$$

32) Suppose that the cost C of renting a car is a linear function of the number m of miles driven. It costs $98.25 to rent a car that was driven 155 miles, and $105.45 to rent a car that was driven 203 miles. Express C as a function of m. Use this function to find the rental cost of a car driven 188 miles. Interpret the the slope and C-intercept of this line.

Answer: $C(m) = 0.15m + 75$, if $m \ge 0$; $C(188) = \$103.20$; The slope is the marginal cost of driving one more mile. The C-intercept is the fixed cost of renting the car.

1.3 Polynomials and Algebraic Functions
Solve the problem.

1) Given $f(x) = \sqrt{x}$ and $g(x) = x - 1$, find the formula and the domain of a) $f + g$, b) $f - g$, c) $f \cdot g$, and d) $\dfrac{f}{g}$.

   Answer: a) $x + \sqrt{x} - 1, x \geq 0$ b) $1 + \sqrt{x} - x, x \geq 0$ c) $\sqrt{x}\,(x - 1), x \geq 0$ d) $\dfrac{\sqrt{x}}{x - 1}, 0 \leq x < 1, x > 1$

2) Given $f(x) = \dfrac{x - 2}{x - 3}$ and $g(x) = \dfrac{x + 2}{x + 3}$, find the formula and the domain of a) $f + g$, b) $f - g$, c) $f \cdot g$, and d) $\dfrac{f}{g}$.

   Answer: a) $\dfrac{2x^2 - 12}{x^2 - 9}, x \neq -3, 3$ b) $\dfrac{2x}{x^2 - 9}, x \neq -3, 3$ c) $\dfrac{x^2 - 4}{x^2 - 9}, x \neq -3, 3$ d) $\dfrac{x^2 + x - 6}{x^2 - x - 6}, x \neq -2, 3$

3) Match the four polynomials to their respective graphs, using the degree of the polynomials, their number of zeros, and their behavior for large |x|.

1) $f(x) = x^5 - x$
2) $f(x) = 2x^5 - 9x^3 + 5x - 3$
3) $f(x) = x^4 - 3x^3 + 6x - 2$
4) $f(x) = 1 + 4x - x^3$

A)

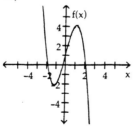

B)

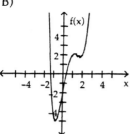

C)

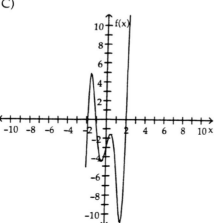

D)

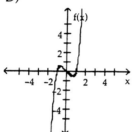

Answer: 1D, 2C, 3B, 4A

Calculus 18

4) Use the vertical limits of the four rational functions to match each with its respective graph.

1) $f(x) = \dfrac{12}{x^2 + 4}$

2) $f(x) = \dfrac{1}{(x+2)(x-1)}$

3) $f(x) = \dfrac{x^2 + 5}{x^2 - 9}$

4) $f(x) = \dfrac{x}{x^2 - 16}$

A)
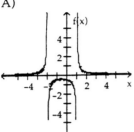

B)
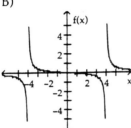

C)
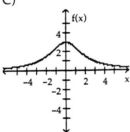

D)
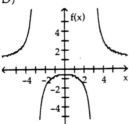

Answer: 1C, 2A, 3D, 4B

5) Use the the domains of the functions to match each function with its respective graph.

1) $f(x) = \sqrt[3]{x^2 - 5x}$
2) $f(x) = \sqrt{x^2 - 5x}$
3) $f(x) = x\sqrt{x+5}$
4) $f(x) = \sqrt{5x - x^2}$

A)

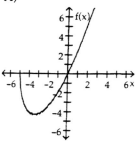

B)

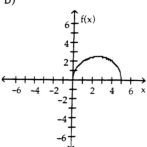

C)

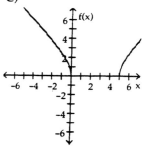

D)

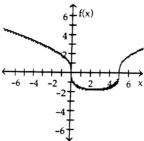

Answer: 1D, 2C, 3A, 4B

6) Use a graphing utility to graph the principle features of $f(x) = x^3 - 4x + 2$. Use the graph to determine the real solutions (to the nearest tenth) of the equation $f(x) = 0$.

Answer:

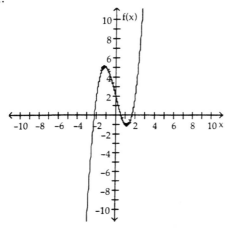

real solutions at $x \approx -2.2, 0.5,$ and 1.7

7) Use a graphing utility to graph the principle features of $f(x) = 2x^4 - 7x^3 + 11x - 4$. Use the graph to determine the real solutions (to the nearest tenth) of the equation $f(x) = 0$.

Answer:

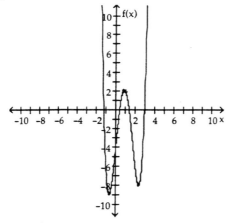

real solutions at $x \approx -1.2, 0.4, 1.4,$ and 2.9

8) Using a graphing utility, determine how the graph of $f(x) = x^3 - 2x + c$ changes on the interval $-4 \leq c \leq 4$.

Answer: The graph moves up the y-axis when c is positive and down the y-axis when c is negative. The y-intercept of the graph is c.

9) Using a graphing utility, determine how the graph of $f(x) = \dfrac{x}{\sqrt{c^2 - x^2}}$ changes on the interval $1 \leq c \leq 10$, x in $(-c, c)$.

Answer: The graph gets wider as c increases. The domain of the function is $-c \leq x \leq c$, and there are vertical asymptotes at $x = -c$ and c.

1.4 Transcendental Functions
Solve the problem.

1) Choose the graph of the function $f(x) = 3^x - 2$ without using a graphing utility.

A)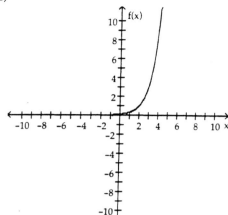

B)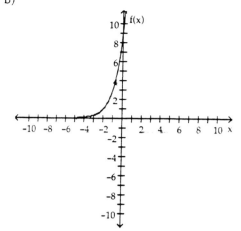

C)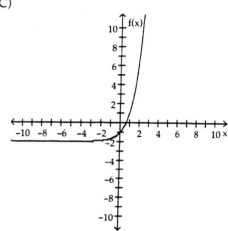

D)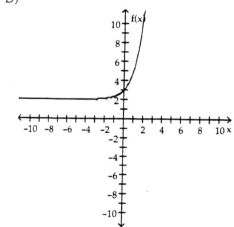

Answer: C

2) Choose the graph of the function $f(x) = 1 - 2^{-x}$ without using a graphing utility.

A)
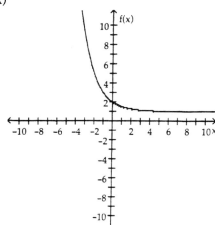

B)
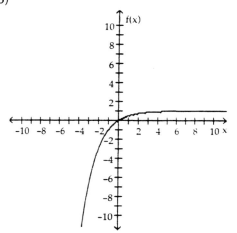

C)
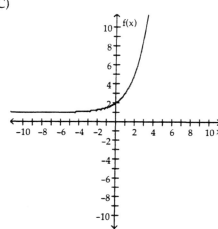

D)
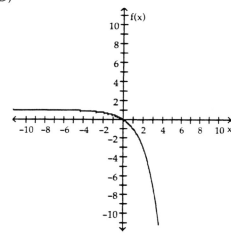

Answer: B

3) Choose the graph of the function $f(x) = 2 + \cos x$ without using a graphing utility.

A)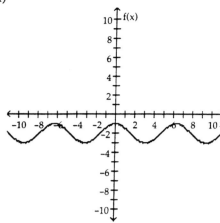

B)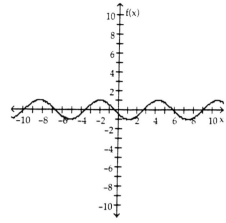

C)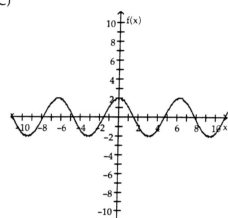

D)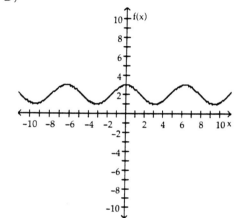

Answer: D

4) Choose the graph of the function $f(x) = 1 - 2\sin x$ without using a graphing utility.

A)
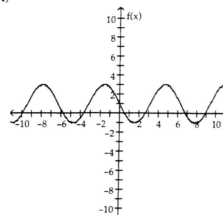

B)
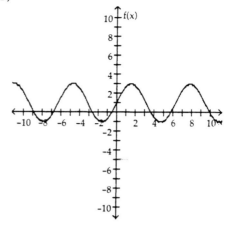

C)
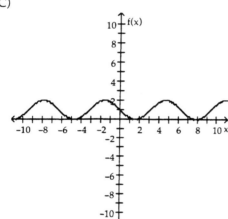

D)
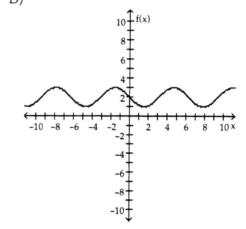

Answer: A

5) Choose the graph of the function $f(x) = \dfrac{x}{3^x}$ without using a graphing utility.

A)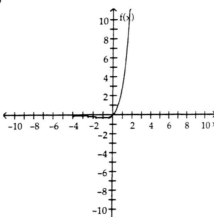

B)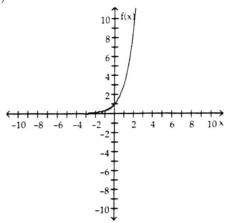

C)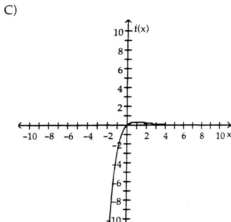

D)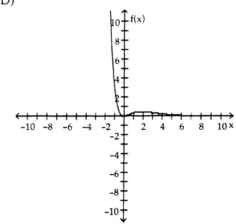

Answer: C

6) Choose the graph of the function $f(x) = \dfrac{\ln x}{x}$ without using a graphing utility.

A)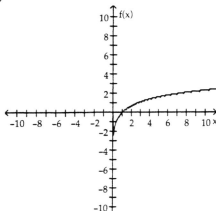

B)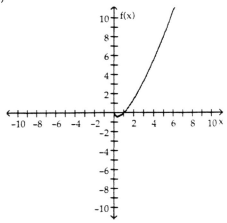

C)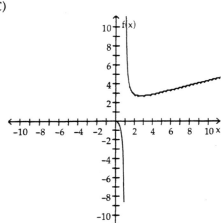

D)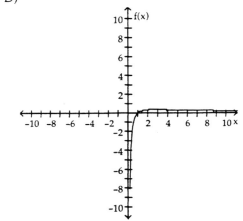

Answer: D

7) Choose the graph of the function $f(x) = \dfrac{1 + \sin 6x}{1 + x^2}$ without using a graphing utility.

A)
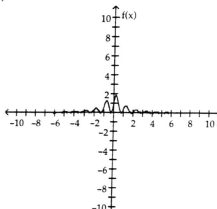

B)
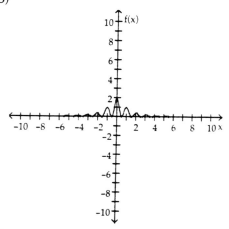

C)
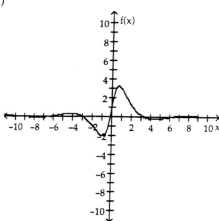

D)
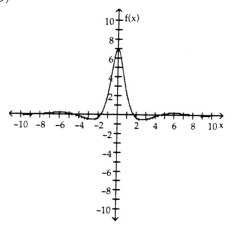

Answer: A

8) If $f(x) = 2 - x^2$ and $g(x) = 4x - 1$, find $f(g(x))$ and $g(f(x))$.

Answer: $f(g(x)) = 1 + 8x - 16x^2$; $g(f(x)) = 7 - 4x^2$

9) If $f(x) = x^2 + 2$ and $g(x) = \dfrac{1}{x^2 + 2}$, find $f(g(x))$ and $g(f(x))$.

Answer: $f(g(x)) = \dfrac{2x^4 + 8x^2 + 9}{x^4 + 4x^2 + 4}$; $g(f(x)) = \dfrac{1}{x^4 + 4x^2 + 6}$

10) If $f(x) = x^3 - 1$ and $g(x) = \sqrt[3]{x + 1}$, find $f(g(x))$ and $g(f(x))$.

Answer: $f(g(x)) = x$; $g(f(x)) = x$

11) If $f(x) = \cos x$ and $g(x) = x^4$, find $f(g(x))$ and $g(f(x))$.

Answer: $f(g(x)) = \cos(x^4)$; $g(f(x)) = \cos^4 x$

12) If $f(x) = \sin x$ and $g(x) = 1 - x^2$, find $f(g(x))$ and $g(f(x))$.

Answer: $f(g(x)) = \sin(1 - x^2)$; $g(f(x)) = 1 - \sin^2 x = \cos^2 x$

13) Find a function $f(x) = x^k$ and a function g such that $f(g(x)) = h(x) = (4x + 5)^3$.

   Answer: $f(x) = x^3$ and $g(x) = 4x + 5$

14) Find a function $f(x) = x^k$ and a function g such that $f(g(x)) = h(x) = \sqrt{3x - x^2}$.

   Answer: $f(x) = x^{1/2}$ and $g(x) = 3x - x^2$

15) Find a function $f(x) = x^k$ and a function g such that $f(g(x)) = h(x) = \dfrac{1}{x - 12}$.

   Answer: $f(x) = x^{-1}$ and $g(x) = x - 12$

16) Find a function $f(x) = x^k$ and a function g such that $f(g(x)) = h(x) = \dfrac{1}{\sqrt{x + 8}}$.

   Answer: $f(x) = x^{-1/2}$ and $g(x) = x + 8$

17) Use a graphing utility to determine the number of real solutions by inspecting the graph of $3^x = x$.

   Answer: no real solutions

18) Use a graphing utility to determine the number of real solutions by inspecting the graph of $x + 1 = 3\sin x$.

   Answer: 3 real solutions

19) Use a graphing utility to determine the number of real solutions by inspecting the graph of $9\cos x = x$.

   Answer: 5 real solutions

20) Use a graphing utility to determine the number of real solutions by inspecting the graph of $2\ln x = \sin x$, $x > 0$.

   Answer: 1 real solution

21) Use a graphing utility to determine the number of real solutions by inspecting the graph of $x^2 = 8\sin x$.

   Answer: 2 real solutions

22) Use a graphing utility to determine the number of real solutions by inspecting the graph of $x = 4\cos x + 10\log_{10} x$, $x > 0$.

   Answer: 6 real solutions

23) In 1990, the population P of a certain species of bird was 52.8 million and growing at a rate of 3.1% per year. If the population continues to grow at this rate, then t years after 1990 it will be $P(t) = 52.8(1.031)^t$ (millions). Determine graphically how long it will take the population of this bird species to double.

   Answer: about 23 years

24) Suppose the population P of an endangered fish decreases at a rate of 0.19% per year. Then after t years the population would be $P(t) = A_0(0.9981)^t$, where A_0 denotes the initial amount. Determine graphically how long it will take for only half of the fish population to be left. Does the numerical value of A_0 affect this answer?

   Answer: about 365 years; no

1.5 Preview: What Is Calculus?

1) There are no exercises for this section.

   Answer:

Ch. 2 Graphs of Equations and Functions
2.1 Tangent Lines and Slope Predictors
Solve the problem.

1) Apply the slope predictor formula to find the slope of the line tangent to $y = f(x) = 3x^2 - 6x$. Then write the equations of the lines tangent to and normal to the graph of f at the point (2, 0).

   Answer: $m(a) = 6$; tangent: $y = 6x - 12$; normal: $y = -\frac{1}{6}x + \frac{1}{3}$

2) Apply the slope predictor formula to find the slope of the line tangent to $y = f(x) = x^2$. Then write the equations of the lines tangent to and normal to the graph of f at the point (3, 9).

   Answer: $m(a) = 2a$; tangent: $y = 6x - 9$; normal: $y = -\frac{1}{6}x + \frac{19}{2}$

3) Apply the slope predictor formula to find the slope of the line tangent to $y = f(x) = x^2 - 3$. Then write the equations of the lines tangent to and normal to the graph of f at the point (2, 8).

   Answer: $m(a) = 2a$; tangent: $y = 16x - 24$; normal: $y = -\frac{1}{16}x + \frac{9}{8}$

4) Apply the slope-predictor formula to find the slope of the line tangent to $y = f(x) = 4 - 3x^2$. Then write the equation of the line tangent to the graph f at the point $(3, f(3))$.

   Answer: $m(a) = -6a$; $y = -18x + 31$

5) Apply the slope predictor formula to find the slope of the line tangent to $y = f(x) = 3x^2 - 5x + 6$. Then write the equation of the line tangent to the graph of f at the point $(3, f(3))$.

   Answer: $m(a) = 6a - 5$; $y = 13x - 21$

6) Apply the slope predictor formula to find the slope of the line tangent to $y = f(x) = 3x - (\frac{x}{2})^2$. Then write the equation of the line tangent to the graph of f at the point $(3, f(3))$.

   Answer: $m(a) = 3 - \frac{a}{2}$; $y = \frac{3}{2}x + \frac{9}{4}$

7) Apply the slope predictor formula to find the slope of the line tangent to $y = f(x) = 3 - (2x + 1)^2$. Then write the equation of the line tangent to the graph of f at the point $(3, f(3))$.

   Answer: $m(a) = -8a - 4$; $y = -28x + 38$

8) Apply the slope predictor formula to find the slope of the line tangent to $y = f(x) = (x + 2)^2 - 3x$. Then write the equation of the line tangent to the graph of f at the point $(3, f(3))$.

   Answer: $m(a) = 2a + 1$; $y = 7x - 5$

9) Apply the slope predictor formula to find the slope of the line tangent to $y = f(x) = (2x + 4)^2 - (2x - 4)^2$. Then write the equation of the line tangent to the graph of f at the point $(3, f(3))$.

   Answer: $m(a) = 32$; $y = 32x$

10) Find all points on the graph of $y = x^2 - x$ at which the tangent line is horizontal.

    Answer: $(\frac{1}{2}, -\frac{1}{4})$

11) Find all points on the curve $y = x^2 - 3x + 1$ at which the tangent line is horizontal.

   Answer: $(\frac{3}{2}, -\frac{5}{4})$

12) Find all points on the curve $y = 3x^2 - 12x - 4$ at which the tangent line is horizontal.

   Answer: $(2, -16)$

13) Find all points on the curve $y = 12x - 2x^2$ at which the tangent line is horizontal.

   Answer: $(3, 18)$

14) Find all points on the curve $y = x - (\frac{x}{10})^2$ at which the tangent line is horizontal.

   Answer: $(50, 25)$

15) Find all points on the curve $y = (x + 4)(x - 5)$ at which the tangent line is horizontal.

   Answer: $(\frac{1}{2}, -\frac{81}{4})$

16) Find all points on the curve $y = 25(1 - \frac{x}{5})^2$ at which the tangent line is horizontal.

   Answer: $(5, 0)$

17) If a ball is thrown straight upward with initial velocity 128 ft/s, then its height t seconds later is $y(t) = 128t - 16t^2$ feet. Determine the maximum height the ball attains by finding the point on the parabola $y(t) = 128t - 16t^2$ where the tangent line is horizontal.

   Answer: 256 feet

18) Find the maximum possible value of the product of two positive numbers whose sum is 100.

   Answer: 2500

19) Suppose that a projectile is fired at an angle of 45° from the horizontal. Its initial position is the origin in the xy-plane, and its initial velocity is $100\sqrt{2}$ ft/s. Then its trajectory will be the part of the parabola $y = x - (\frac{x}{25})^2$ for which $y \geq 0$. (a) How far does the projectile travel (horizontally) before it hits the ground?

   (b) What is the maximum height above the ground that the projectile attains?

   Answer: (a) 625 feet; (b) 156.25 feet

20) One of the two lines that pass through the point (2, 0) and are tangent to the parabola $y = x^2$ is the x-axis. Find an equation for the *other* line. *(Suggestion:* First find the coordinates of the number *a* shown on the graph).

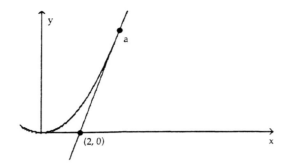

Answer: $y = 8x - 16$

2.2 The Limit Concept
Solve the problem.

1) Evaluate $\lim_{x \to 2} \dfrac{\frac{1}{x} - \frac{1}{2}}{x - 2}$.

Answer: $-\dfrac{1}{4}$

2) Evaluate $\lim_{h \to 0} \dfrac{(3 + h)^4 - 81}{h}$.

Answer: 108

3) Evaluate $\lim_{h \to 0} \dfrac{\sqrt{x + h + 5} - \sqrt{x + 5}}{h}$.

Answer: $\dfrac{1}{2\sqrt{x + 5}}$

4) Evaluate $\lim_{h \to 0} \dfrac{\frac{1}{8 + 3h} - \frac{1}{8}}{h}$.

Answer: $-\dfrac{3}{64}$

5) Evaluate $\lim_{h \to 0} \dfrac{[2(x + h) + 3]^3 - [2x + 3]^3}{h}$.

Answer: $6(2x + 3)^2$

6) Evaluate $\lim_{x \to 5} \dfrac{x^2 - 25}{x - 5}$.

Answer: 10

Calculus 32

7) Evaluate $\lim_{x \to 4} \frac{x-4}{\sqrt{x}-2}$.

Answer: 4

8) Evaluate $\lim_{x \to 9} \frac{\sqrt{x}-3}{x-9}$.

Answer: $\frac{1}{6}$

9) Evaluate $\lim_{x \to 3} \frac{2x-6}{x^2-4x+3}$.

Answer: 1

10) Evaluate $\lim_{x \to 3} \frac{x^2-x-6}{x-3}$.

Answer: 5

11) Evaluate $\lim_{x \to 16} \frac{\sqrt{x}-4}{x-16}$.

Answer: $\frac{1}{8}$

12) Evaluate $\lim_{x \to 1} \frac{x^4-1}{x^3-1}$.

Answer: $\frac{4}{3}$

13) Given $f(x) = 3x^2$, use the four step process to find a slope-predictor function $m(x)$. Then write an equation for the line tangent to the curve at the point $x = 2$.

Answer: $m(x) = 6x$; $y = 12x - 12$

14) Given $f(x) = \frac{2}{x^2}$, use the four step process to find a slope-predictor function $m(x)$. Then write an equation for the line tangent to the curve at the point $x = 1$.

Answer: $m(x) = \frac{-4}{x^3}$; $y = -4x + 6$

15) Given $f(x) = \frac{2}{x-1}$, use the four step process to find a slope-predictor function $m(x)$. Then write an equation for the line tangent to the curve at the point $x = 0$.

Answer: $m(x) = \frac{-2}{(x-1)^2}$; $y = -2x - 2$

16) Given $f(x) = \dfrac{4}{\sqrt{x+8}}$, use the four step process to find a slope-predictor function $m(x)$. Then write an equation for the line tangent to the curve at the point $x = 8$.

Answer: $m(x) = \dfrac{-2}{(\sqrt{x+8})^3}$; $y = \dfrac{-x}{32} + \dfrac{5}{4}$

17) Given $f(x) = \dfrac{2x^2}{x-1}$, use the four step process to find a slope-predictor function $m(x)$. Then write an equation for the line tangent to the curve at the point $x = 2$.

Answer: $m(x) = \dfrac{2x(x-2)}{(x-1)^2}$; $y = 8$

2.3 More About Limits
Solve the problem.

1) Find the trigonometric limit: $\lim\limits_{\theta \to 0} \dfrac{\theta^2}{\cos\theta}$.

Answer: 0

2) Find the trigonometric limit: $\lim\limits_{\theta \to 0} \dfrac{\tan\theta}{\theta}$.

Answer: 1

3) Find the trigonometric limit: $\lim\limits_{x \to 0} \dfrac{\sin 3x}{2x}$.

Answer: $\dfrac{3}{2}$

4) Find the trigonometric limit: $\lim\limits_{x \to 0} \dfrac{\sin 2x}{\sqrt{2x}}$.

Answer: 0

5) Find the trigonometric limit: $\lim\limits_{x \to 0} \dfrac{1 - \sin 2x}{x}$.

Answer: -1

6) Find the trigonometric limit: $\lim\limits_{\theta \to 0} \dfrac{1 - \sin\theta}{\theta \sin\theta}$.

Answer: 1

7) Find the trigonometric limit: $\lim\limits_{x \to 0} \dfrac{x - \tan 2x}{\sin 2x}$.

Answer: $-\dfrac{1}{2}$

8) Find the trigonometric limit: $\lim\limits_{x \to 0} \dfrac{\sin 4x}{\sin 5x}$.

Answer: $\dfrac{4}{5}$

9) Find the trigonometric limit: $\lim_{\theta \to 0} \dfrac{2\sin 2\theta}{3\theta}$.

   Answer: 0

10) Use the squeeze law of limits to find the limit.
    $$\lim_{x \to 0} x^2 \sin^2 10x$$

    Answer: 0

11) Use the squeeze law of limits to find the limit.
    $$\lim_{x \to 0} (\sqrt[3]{x})(\cos \tfrac{1}{x})$$

    Answer: 0

12) Use one-sided limits to find the limit or determine that the limit does not exist.
    $$\lim_{x \to 9^+} (3 - \sqrt{x})$$

    Answer: 0

13) Use one-sided limits to find the limit or determine that the limit does not exist.
    $$\lim_{x \to 3^+} \sqrt{x^2 - 9}$$

    Answer: 0

14) Use one-sided limits to find the limit or determine that the limit does not exist.
    $$\lim_{x \to 2^+} \sqrt{\dfrac{2x}{x-2}}$$

    Answer: $+\infty$ (does not exist)

15) Use one-sided limits to find the limit or determine that the limit does not exist.
    $$\lim_{x \to 2^+} \sqrt{\dfrac{x^2 - 4x + 4}{x - 2}}$$

    Answer: 0

16) Use one-sided limits to find the limit or determine that the limit does not exist.
    $$\lim_{x \to 4^+} \dfrac{16 - x^2}{4 - x}$$

    Answer: 8

17) Given $f(x) = \dfrac{2}{x - 2}$, find the point where both the right and left-hand limits fail to exist. Describe the behavior of $f(x)$ near this point.

    Answer: $x = 2$; to the right of $x = 2$, $f(x)$ increases without bound and to the left of $x = 2$, $f(x)$ decreases without bound

18) Given $f(x) = \dfrac{|3-x|}{(3-x)^2}$, find the point where both the right and left-hand limits fail to exist. Describe the behavior of $f(x)$ near this point.

Answer: $x = 3$; to the right of $x = 3$, $f(x)$ increases without bound and to the left of $x = 3$, $f(x)$ decreases without bound

19) Given $f(x) = \dfrac{x-3}{x^2 - 6x + 9}$, find the point where both the right and left-hand limits fail to exist. Describe the behavior of $f(x)$ near this point.

Answer: $x = 3$; to the right of $x = 3$, $f(x)$ increases without bound and to the left of $x = 3$, $f(x)$ decreases without bound

20) Given $f(x) = \dfrac{x^2 - 9}{|x - 3|}$, (a) find the left and right-hand limits at $x = 3$. (b) Does the 2-sided limit exist at $x = 3$? (c) Sketch the graph of $y = f(x)$.

Answer: (a) left-hand limit = -6; right-hand limit = +6; (b) no;

(c)
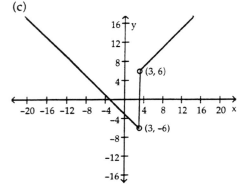

21) Given $f(x) = \dfrac{x^4 - 10x + 25}{|x - 5|}$, (a) find the left and right-hand limits at $x = 5$. (b) Does the 2-sided limit exist at $x = 5$? (c) Sketch the graph of $y = f(x)$.

Answer: (a) left-hand limit = $+\infty$; right-hand limit = $+\infty$; (b) no;

(c)
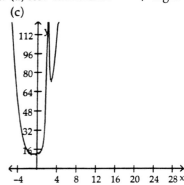

2.4 The Concept of Continuity
Solve the problem.

1) Apply the limit laws and theorems to show that $f(x) = 3x^5 - 4x^2 + 6$ is continuous for all x.

   Answer: $f(a) = 3a^5 - 4a^2 + 6$ is defined for all $a \in R$

   $\lim_{x \to a} 3x^5 - 4x^2 + 6$ exists and $= f(a)$ for all $a \in R$

   therefore, f is continuous

2) Apply the limit laws and theorems to show that $g(x) = \dfrac{3x - 2}{6x^2 + 5}$ is continuous for all x.

   Answer: $6x^2 + 5 \neq 0$ for all $x \in R$

   $g(a) = \dfrac{3a - 2}{6a^2 + 5}$ is defined for all $a \in R$

   $\lim_{x \to a} \dfrac{3x - 2}{6x^2 + 5}$ exists and $= g(a)$ for all $a \in R$

   therefore, g is continuous for all x

3) Apply the limit laws and theorems to show that $g(x) = \dfrac{x^3}{x^2 + 4x + 1}$ is continuous for all x.

   Answer: $x^2 + 4x + 1 \neq 0$ for all $x \in R$

   $g(a) = \dfrac{a^3}{a^2 + 4a + 1}$ is defined for all $a \in R$

   $\lim_{x \to a} \dfrac{x^3}{x^2 + 4x + 1}$ exists and $= g(a)$ for all $a \in R$

   therefore, g is continuous for all x

4) Apply the limit laws and theorems to show that $f(x) = \dfrac{1 - \cos x}{1 + \sin^2 x}$ is continuous for all x.

   Answer: $1 + \sin^2 x \neq 0$ for all $x \in R$

   $f(a) = \dfrac{1 - \cos x}{1 + \sin^2 x}$ is defined for all $a \in R$

   $\lim_{x \to a} \dfrac{1 - \cos x}{1 + \sin^2 x}$ exists and $= f(a)$ for all $a \in R$

   therefore, f is continuous for all x

5) Apply the limit laws and theorems to show that $g(x) = \sqrt{1 - \cos^2 x}$ is continuous for all x.

   Answer: $1 - \cos^2 x = \sin^2 x \geq 0$ for all $x \in R$

   $g(x) = \sqrt{\sin^2 x} = \sin x$ is defined for all $x \in R$

   $g(a) = \sin a$ is defined for all $a \in R$

   $\lim_{x \to a} \sin a$ exists and $= g(a)$ for all $a \in R$

   therefore, g is continuous for all x

6) Given $f(x) = 2x^5 - 7x^2 + 13$, is f continuous for all values of x? Why or why not?

   Answer: yes; f is a polynomial

7) Given $f(x) = \dfrac{3x - 4}{5x^2 + 2}$, is f continuous for all values of x? Why or why not?

Answer: yes; f is a rational function and there are no real values of x which make the denominator = 0.

8) Given $f(x) = \dfrac{1 - \cos x}{1 + \sin^2 x}$, is f continuous for all values of x? Why or why not?

Answer: yes; $1 + \sin^2 x \neq 0$ for any real value of x.

9) Given $f(x) = \dfrac{x^2}{x^3 + 2x^2 + 5x}$, is f continuous for all values of x? Why or why not?

Answer: yes; $x^3 + 2x^2 + 5x \neq 0$ for all real values of x.

10) Given $f(x) = \sqrt[4]{1 - \cos^2 x}$, is f continuous for all values of x? Why or why not?

Answer: yes; $1 - \cos^2 x = \sin^2 x$ and $\sin^2 x \geq 0$ for all real x.

11) Given $f(x) = \dfrac{1}{x + 3}$, show that that f is continuous on the interval $x > -3$

Answer: discontinuous at $x = -3$ which is not in the interval so f is continuous on $x > -3$

12) Given $f(x) = \dfrac{x - 2}{x^2 - 9}$, show that that f is continuous on the interval $-3 < x < 3$

Answer: discontinuous at $x = \pm 3$ which are not in the interval so f is continuous on $-3 < x < 3$

13) Given $f(x) = \dfrac{x}{\sin x}$, show that that f is continuous on the interval $-\pi < x < \pi$

Answer: discontinuous at $x = \pm \pi$, which are not in the interval, so f is continuous on $-\pi < x < \pi$

14) Given $f(x) = x^3 + \dfrac{1}{x}$, tell where f is continuous. (Give your answer in interval form.)

Answer: $(-\infty, 0) \cup (0, +\infty)$

15) Given $g(x) = \dfrac{3}{x - 3}$, tell where g is continuous. (Give your answer in interval form.)

Answer: $(-\infty, 3) \cup (3, +\infty)$

16) Given $h(x) = \dfrac{x - 9}{|x - 9|}$, tell where h is continuous. (Give your answer in interval form.)

Answer: $(-\infty, 9) \cup (9, +\infty)$

17) Given $G(x) = \dfrac{2x^2 + 5}{2x - 10}$, tell where G is continuous. (Give your answer in interval form.)

Answer: $(-\infty, 5) \cup (5, +\infty)$

18) Given $F(x) = \dfrac{2x^2 + 3x + 6}{x^2 + 3}$, tell where F is continuous. (Give your answer in interval form.)

Answer: $(-\infty, +\infty)$

19) Given $f(x) = \sqrt{\dfrac{2x+3}{2x-3}}$, tell where f is continuous. (Give your answer in interval form.)

Answer: $(-\infty, -\dfrac{3}{2}) \cup (\dfrac{3}{2}, +\infty)$

20) Given $f(x) = \sqrt{\dfrac{4-x^2}{16-x^2}}$, tell where f is continuous. (give your answer in interval form)

Answer: $(-\infty, -4) \cup [-2, 2] \cup (4, +\infty)$

21) Given $f(x) = \dfrac{\cos x}{x^3}$, tell where f is continuous. (Give your answer in interval form.)

Answer: $(-\infty, 0) \cup (0, +\infty)$

22) Given $L(x) = \dfrac{x}{\sin x}$, tell where L is continuous. (Give your answer in interval form.)

Answer: continuous for all real numbers except $x = 0$ and integral multiples of π

23) Given $H(x) = 3\cos|2x|$, tell where H is continuous. (Give your answer in interval form.)

Answer: $(-\infty, +\infty)$

24) Given $f(x) = \dfrac{1}{\sqrt{1+\sin x}}$, tell where f is continuous. (Give your answer in interval form.)

Answer: continuous for all real numbers except odd multiples of $\dfrac{\pi}{2}$

25) Given $f(x) = \dfrac{x-4}{x^2-16}$, find all points where f is not defined (and therefore not continuous). For each such point, tell whether or not the discontinuity is removable.

Answer: $x = 4$ is removable; $x = -4$ is non-removable

26) Given $f(x) = \dfrac{4}{4-|2x|}$, find all points where f is not defined (and therefore not continuous). For each such point, tell whether or not the discontinuity is removable.

Answer: $x = \pm 2$ is non-removable

27) Given $f(x) = \begin{cases} 2x, & x > 0 \\ -x^2, & x < 0 \end{cases}$, find all points where f is not defined (and therefore not continuous). For each such point, tell whether or not the discontinuity is removable.

Answer: $x = 0$ is removable

28) Given $f(x) = \begin{cases} 2x + 4, & x > \frac{2}{7} \\ 6 - 5x, & x \leq \frac{2}{7} \end{cases}$, find all points where f is not defined (and therefore not continuous). For each such point, tell whether or not the discontinuity is removable.

Answer: continuous on $(-\infty, +\infty)$

29) Given $f(x) = \begin{cases} x^2, & x \geq 0 \\ \frac{\cos x}{x^2}, & x < 0 \end{cases}$, find all points where f is not defined (and therefore not continuous). For each such point, tell whether or not the discontinuity is removable.

Answer: $x = 0$ is non-removable

30) Given $f(x) = \begin{cases} \frac{1 - \sin x}{2x}, & x \leq 0 \\ x^2 - 4, & x > 0 \end{cases}$, find all points where f is not defined (and therefore not continuous). For each such point, tell whether or not the discontinuity is removable.

Answer: $x = 0$ removable

31) Given $f(x) = \begin{cases} 2 - x^2, & \text{if } x \geq 0 \\ x + c, & \text{if } x < 0 \end{cases}$, find a value for c so that $f(x)$ is continuous for all x.

Answer: $c = 2$

32) Given $f(x) = \begin{cases} c^2 - x^2, & \text{if } x < 0 \\ c \cos x, & \text{if } x \geq 0 \end{cases}$, find a value for c so that $f(x)$ is continuous for all x.

Answer: $c = 1$

33) Given $x^3 - 6x + 4 = 0$, show that the equation has 3 distinct real roots by calculating the values of the left-hand side at $x = -3, -2, -1, 0, 1, 2, 3$. Then apply the intermediate value property of continuous functions on appropriate closed intervals.

Answer: $f(-3) = -5$, $f(-2) = 8$, $f(-1) = 9$, $f(0) = 4$, $f(1) = -1$, $f(2) = 0$, $f(3) = 13$
f is continuous on $[-3, -2]$, $[-2, 1]$, $[1, 2]$

34) Given $x^3 - 6x - 2 = 0$, show that the equation has 3 distinct real roots by calculating the values of the left-hand side at $x = -3, -2, -1, 0, 1, 2, 3$. Then apply the intermediate value property of continuous functions on appropriate closed intervals.

Answer: $f(-3) = -11$, $f(-2) = 2$, $f(-1) = 3$, $f(0) = -2$, $f(1) = -7$, $f(2) = -6$, $f(3) = 7$
f is continuous on $[-3, -2]$, $[-1, 0]$, $[2, 3]$

35) Determine where the function $f(x) = x + [\![x^2]\!] - [\![x]\!]$ is continuous.

Answer: $...(-3, -2), (-2, -1), (-1, 2), (2, 3), (3, 4),...$

36)

Given $f(x) = 4^{-2/x^2}$, $x \neq 0$; $f(0) = 0$. Is f continuous?

Answer: yes; The only point of discontinuity is $x = 0$ which is a removable discontinuity.

37)

Given $f(x) = \left(\dfrac{1}{1 + 2^{-1/x}} \right)$, $x \neq 0$; $f(0) = 1$. Is f continuous?

Answer: no; The point of discontinuity, $x = 0$, is not a removable discontinuity.

Ch. 3 The Derivative

3.1 The Derivative and Rates of Change

Solve the problem.

1) Find $f'(x)$ given $f(x) = 3x + 9$ by using the differentiation rule: If $f(x) = ax^2 + bx + c$, then $f'(x) = 2ax + b$.

   Answer: $f'(x) = 3$

2) Find $h'(z)$ given $h(z) = 3z(9 - z)$ by using the differentiation rule: If $f(x) = ax^2 + bx + c$, then $f'(x) = 2ax + b$.

   Answer: $h'(z) = 27 - 6z$

3) Find dy/dx given $y = 5x^2 + 3x - 17$ by using the differentiation rule: If $f(x) = ax^2 + bx + c$, then $f'(x) = 2ax + b$.

   Answer: $dy/dx = 10x + 3$

4) Find du/dt given $u = -3t^2 - 7t - 1$ by using the differentiation rule: If $f(x) = ax^2 + bx + c$, then $f'(x) = 2ax + b$.

   Answer: $du/dt = -6t - 7$

5) Use the definition of the derivative to find $f'(x)$ given $f(x) = 3x + 2$.

   Answer: $f'(x) = 3$

6) Use the definition of the derivative to find $f'(x)$ given $f(x) = x^2 - 1$.

   Answer: $2x$

7) Use the definition of the derivative to find $g'(x)$ given $g(x) = (x - 1)^2$.

   Answer: $2(x - 1)$

8) Use the definition of the derivative to find $f'(x)$ given $f(x) = x^2 + x + 1$.

   Answer: $2x + 1$

9) Use the definition of the derivative to find $f'(x)$ given $f(x) = \sqrt{x}$.

   Answer: $f'(x) = \dfrac{1}{2\sqrt{x}}$

10) Use the definition of the derivative to find $f'(x)$ given $f(x) = \sqrt{3x + 1}$.

    Answer: $f'(x) = \dfrac{3}{2\sqrt{3x + 1}}$

11) Use the definition of the derivative to find $g'(x)$ given $g(x) = x^3 - 1$.

    Answer: $3x^2$

12) Use the definition of the derivative to find $f'(x)$ given $f(x) = \dfrac{x}{x + 1}$.

    Answer: $\dfrac{1}{(x + 1)^2}$

13) A particle moves along the x-axis; at time t its location is $x(t) = 3t^2 - 6t + 1$. Find its position when its velocity is zero.

    Answer: -2

14) A ball thrown vertically upward at time $t = 0$ (seconds) has height $y(t) = 96t - 16t^2$ (ft) at time t. Find the maximum height that the ball attains.

Answer: 144 ft

15) Match the graph of the function to the graph of its derivative.

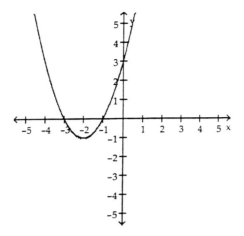

A)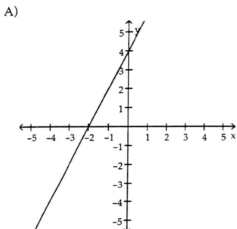

B)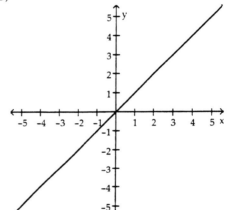

C)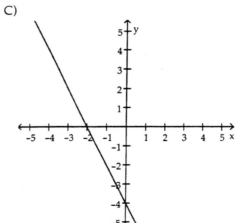

D)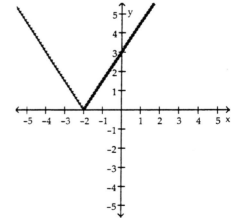

Answer: A

Calculus 44

16) Match the graph of the function with the graph of its derivative.

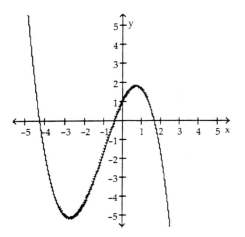

A)
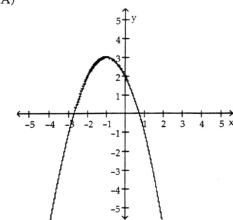

B)
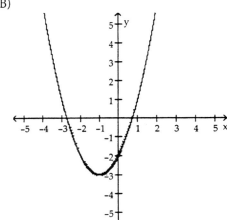

C)
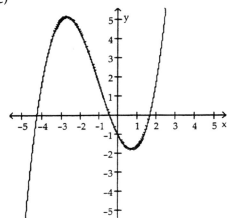

D)
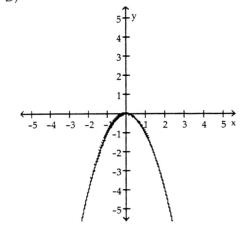

Answer: A

17) Match the graph of the function to the graph of its derivative.

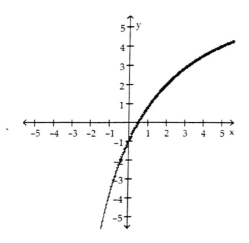

A)
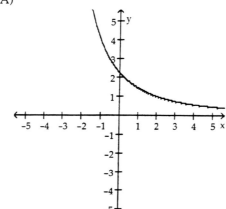

B)
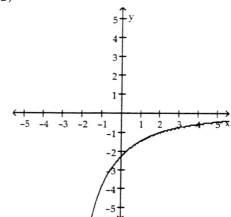

C)
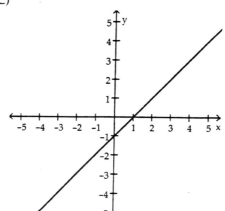

D)
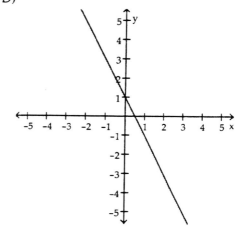

Answer: A

18) Find the rate of change of the volume V of a cube with respect to the length x of one edge.

Answer: $V' = 3x^2$

19) Find the rate of change of the surface area A of a cube with respect to the length x of one edge.

Answer: $A' = 12x$

20) At time $t = 0$ (seconds), a race-car begins to speed down a track; it travels $10t^2$ feet in the first t seconds. What is its velocity when $t = 10$?

Answer: 200

21) Water leaks from a tank in such a way that the volume of water in the tank at the time t (minutes) is $(80 - t)^2$ (liters). Find the instantaneous rate of change of water in the tank at time $t = 40$.

Answer: −80 liters/min

22) You drop a ball from a cliff above the sea. Its height above the ocean, in feet, t seconds after you throw it is $s(t) = -16t^2 + 16t + 320$. (a) Find the average velocity of the ball from time $t = 2$ to time $t = 2 + h$. (b) Find the instantaneous velocity of the ball at time $t = 2$.

Answer: (a) $-16h - 48$ ft/s
(b) -48 ft/s

23) A ball thrown vertically upward at time $t = 0$ (s) with initial velocity 80 ft/s and with initial height 96 ft has height function $y(t) = -16t^2 + 80t + 96$. (a) What is the maximum height attained by the ball? (b) When and with what impact speed does the ball hit the ground?

Answer: (a) 196 ft, (b) when $t = 6$ s the ball hits the ground with impact speed 112 ft/s

3.2 Basic Differentiation Rules
Solve the problem.

1) Find $g'(z)$ given $g(z) =$ a) $z^2 - 5z + 7$, b) $z - \left(\dfrac{z}{4}\right)^2$.

Answer: a) $2z - 5$, b) $1 - \dfrac{z}{8}$

2) Find $g'(x)$ given $g(x) = (x - 1)^2$.

Answer: $2(x - 1)$

3) Find $f'(x)$ given $f(x) = x^2 + x + 1$.

Answer: $2x + 1$

4) Find $g'(x)$ given $g(x) = x^3 - 1$.

Answer: $3x^2$

5) Find $g'(x)$ given $g(x) = x^4$.

Answer: $4x^3$

6) Find $f'(x)$ given $f(x) = \dfrac{x}{x + 1}$.

Answer: $\dfrac{1}{(x + 1)^2}$

7) Find $f'(x)$ given $f(x) =$ (a) $12x^5 - 7x^3 + 7x^2 - 14x + 24$, (b) $\dfrac{x+2}{x+3}$, (c) $(x^5 + x)^2$, (d) x^{-1}.

Answer: (a) $60x^4 - 21x^2 + 14x - 14$, (b) $\dfrac{1}{(x+3)^2}$, (c) $2(x^5 + x)(5x^4 + 1)$, (d) $-\dfrac{1}{x^2}$

8) Find $h'(x)$ given $h(x) = (3x+4)(4x-3)$.

Answer: $24x + 7$

9) Find $g'(x)$ given $g(x) = (x^3 + 1)(3x^2 - 1)$.

Answer: $15x^4 - 3x^2 + 6x$

10) Find $f'(y)$ given $f(y) = 3y(2y+5)(2y-5)$.

Answer: $36y^2 - 75$

11) Find $g'(x)$ given $g(x) = \dfrac{1}{x+2} - \dfrac{1}{x-2}$.

Answer: $\dfrac{-1}{(x+2)^2} + \dfrac{1}{(x-2)^2}$

12) Find $g'(t)$ given $g(t) = (t^2 + 3)(2t^3 + 3t^2 + 5)$.

Answer: $10t^4 + 12t^3 + 18t^2 + 28t$

13) Find $f'(x)$ given $f(x) = \dfrac{5x^3 - 4x^2 + 3x - 2}{x^2}$.

Answer: $5 - \dfrac{3}{x^2} + \dfrac{4}{x^3}$

14) Find $g'(t)$ given $g(t) = \dfrac{t-5}{t^2 + 10t + 25}$.

Answer: $\dfrac{-t^2 + 10t + 75}{(t^2 + 10t + 25)^2} = -\dfrac{t-15}{(t+5)^3}$

15) Find $u'(x)$ given $u(x) = \dfrac{1}{(x-3)^2}$.

Answer: $-\dfrac{2}{(x-3)^3}$

16) Find $h'(x)$ given $h(x) = \dfrac{4x^3 + x^2 - 2x + 7}{4x - 3}$.

Answer: $\dfrac{32x^3 - 32x^2 - 6x - 22}{(4x-3)^2}$

17) Find $y'(x)$ given $y(x) = x^4 - 5x^6 + \dfrac{4}{5}x^{-5} + 7$.

Answer: $4x^3 - 30x^5 - \dfrac{4}{x^6}$

18) Find $y'(x)$ given $y(x) = \dfrac{x}{x+2} + \dfrac{x-2}{5x}$.

   Answer: $\dfrac{2}{(x+2)^2} + \dfrac{2}{5x^2}$

19) Find $w'(z)$ given $w(z) = z^3\left(3z^4 - \dfrac{3}{2z^4}\right)$.

   Answer: $21z^6 + \dfrac{3}{2z^2}$

20) Find $y'(x)$ given $y(x) = \dfrac{3x^2}{2x - \dfrac{3}{5x^3}}$.

   Answer: $\dfrac{150x^8 - 225x^4}{(10x^4 - 3)^2}$

21) Find $h'(w)$ given $h(w) = \dfrac{w+7}{w^3}$.

   Answer: $-\dfrac{2}{w^3} - \dfrac{21}{w^4}$

22) Write an equation of the line tangent to the graph of the function $p(x) = 3x^2 - 6x$ at the point $P(2, 0)$. Express the answer in the form $ax + by = c$.

   Answer: $6x - y = 12$

23) Write an equation of the line tangent to the graph of $y = x^2$ at the point $P(3, 9)$. Express the answer in the form $ax + by = c$.

   Answer: $6x - y = 9$

24) Write an equation of the straight line that is tangent to the graph of $y = x^3$ at the point $P(2, 8)$. Express the answer in the form $ax + by = c$.

   Answer: $12x - y = 16$

25) Write an equation of the straight line tangent to the graph of $f(x) = \dfrac{1}{x}$ at the point $P(1, 1)$. Express the answer in the form $ax + by = c$.

   Answer: $x + y = 2$

26) Write an equation of the straight line tangent to the graph $y = x^5$ at the point $P(2, 32)$. Express the answer in the form $ax + by = c$.

   Answer: $80x - y = 128$

Calculus 49

27) Write an equation of the straight line tangent to the graph of $y = \dfrac{6}{1-x^2}$ at the point $P(2, -2)$. Express the answer in the form $ax + by = c$.

Answer: $8x - 3y = 22$

28) Find the equation of the tangent to the graph of $f(x) = x^2 + 2$ at the point $P(1, 3)$. Express the answer in the form $ax + by = c$.

Answer: $2x - y = -1$

29) Write an equation of the line tangent to the curve $y = \dfrac{5}{x^2} - \dfrac{2}{x^3}$ at the point $P(-1, 7)$. Express the answer in the form $ax + by = c$.

Answer: $16x - y = -23$

30) Write an equation of the line tangent to the curve $y = \dfrac{4x^2}{x^2 + 2x + 1}$ at the point $P(1, 1)$. Express the answer in the form $ax + by = c$.

Answer: $x - y = 0$

31) The function $f(x) = 9\pi x^2$ describes the volume, $f(x)$, of a right circular cylinder of height 9 feet and radius x feet. If the radius is changing, find the rate of change of the volume with respect to the radius when the radius is 10 feet. Leave answer in terms of π.

Answer: 180π ft^3/ft

32) Find two straight lines of slope -3 that are tangent to the curve $y = \dfrac{1}{x}$.

Answer: $3x + y = 2\sqrt{3}$, $3x + y = -2\sqrt{3}$

3.3 The Chain Rule
Solve the problem.

1) Find dy/dx given $y =$ (a) $(x^2 + 1)^5$ (b) $\dfrac{x}{(x^2+1)^5}$

 (c) $(x^5 + x)^{19}$ (d) $(x^4 + 1)^5 (x^5 + 1)^4$

 Answer: (a) $10x(x^2 + 1)^4$

 (b) $\dfrac{1 - 9x^2}{(x^2+1)^6}$

 (c) $19(5x^4 + 1)(x^5 + x)^{18}$

 (d) $20x^4(x^4 + 1)^5(x^5 + 1)^3 + 20x^3(x^4 + 1)^4(x^5 + 1)^4 = 20x^3(x^4 + 1)^4(x^5 + 1)^3[2x^5 + x + 1]$

2) Find dy/dx given $y = \dfrac{1}{(3x - 2)^3}$.

Answer: $-\dfrac{9}{(3x - 2)^4}$

3) Find dy/dx given $y = (x^2 + 5x + 6)^3$.

Answer: $3(x^2 + 5x + 6)^2(2x + 5)$

4) Find dy/dx given $y = (5 - 3x^2)^{-3}$.

   Answer: $-3(5 - 3x^2)^{-4}(-6x) = \dfrac{18x}{(3x^2 - 5)^4}$

5) Find dy/dx given $y = [2 + (3 + x)^2]^3$.

   Answer: $6[2 + (3 + x)^2]^2(3 + x) = 6(x + 3)(x^2 + 6x + 11)^2$

6) Express the derivative dy/dx in terms of x if $y = (u + 1)^5$ and $u = \dfrac{1}{2x^2}$.

   Answer: $-\dfrac{5}{x^3}\left[\dfrac{1}{2x^2} + 1\right]^4$

7) Express the derivative dy/dx in terms of x if $y = (1 + u^3)^4$ and $u = (3x - 2)^2$.

   Answer: $72[1 + (3x - 2)^6]^3(3x - 2)^3$

8) Express the derivative dy/dx in terms of x if $y = u(1 + u)^5$ and $u = \dfrac{1}{x^3}$.

   Answer: $-\dfrac{3(1 + x^{-3})^5}{x^4} - \dfrac{15(1 + x^{-3})^4}{x^7} = -\dfrac{3(x^3 + 1)^4(x^3 + 6)}{x^{19}}$

9) Given $f(x) = (3x - x^2)^5$, identify a function u of x and an integer $n \neq 1$ such that $f(x) = u^n$. Then compute $f'(x)$.

   Answer: $u(x) = 3x - x^2$, $n = 5$; $f'(x) = 5(3x - x^2)^4(3 - 2x)$

10) Given $f(x) = \dfrac{1}{(3 - 2x^4)^3}$, identify a function u of x and an integer $n \neq 1$ such that $f(x) = u^n$. Then compute $f'(x)$.

    Answer: $u(x) = 3 - 2x^4$, $n = -3$; $f'(x) = 24x^3(3 - 2x^4)^{-4} = \dfrac{24x^3}{(2x^4 - 3)^4}$

11) Given $f(x) = \dfrac{(x^2 + 3x + 1)^5}{(x + 3)^5}$, identify a function u of x and an integer $n \neq 1$ such that $f(x) = u^n$. Then compute $f'(x)$.

    Answer: $u = \dfrac{x^2 + 3x + 1}{x + 3}$, $n = 5$; $f'(x) = \dfrac{5(x^2 + 3x + 1)^4(x^2 + 6x + 8)}{(x + 3)^6}$

12) Differentiate the function $g(y) = y + (3y - 5)^6$.

    Answer: $g'(y) = 1 + 18(3y - 5)^5$

13) Differentiate the function $F(s) = \left(3s - \dfrac{1}{s^3}\right)^2$.

    Answer: $F'(s) = 2(3s - s^{-3})(3 + 3s^{-4}) = \dfrac{6(s^4 + 1)(3s^4 - 1)}{s^7}$

14) Differentiate the function $g(w) = (w^2 - 2w + 3)(w + 3)^4$.

    Answer: $g'(w) = (2w - 2)(w + 3)^4 + 4(w + 3)^3(w^2 - 2w + 3) = 2(w + 3)^3(3w^2 - 2w + 3)$

Calculus 51

15) Differentiate the function $F(z) = \dfrac{1}{(5 - 7z + 3z^6)^{11}}$.

Answer: $F'(z) = 11(7 - 18z^5)(5 - 7z + 3z^6)^{-12} = \dfrac{11(7 - 18z^5)}{(5 - 7z + 3z^6)^{12}}$

16) For each function given below, its derivative can be computed in two ways—one way using the chain rule, another way not using the chain rule. Compute the derivative of each function in both these ways and verify that the results are the same.

 (a) $f(x) = (x^2 + 1)^2 = x^4 + 2x^2 + 1$
 (b) $g(x) = (x^3)^4 = x^{12}$
 (c) $h(x) = (x^4 + 1)^{-1} = \dfrac{1}{x^4 + 1}$
 (d) $j(x) = (x + 2)^3 = x^3 + 6x^2 + 12x + 8$
 (e) $p(x) = (2x)^{10} = 1024x^{10}$

Answer: (a) $4x(x^2 + 1)$
 (b) $12x^{11}$
 (c) $\dfrac{-4x^3}{(1 + x^4)^2}$
 (d) $3(x + 2)^2$
 (e) $10240x^9$

17) Each edge x of a square is increasing at the rate of 3 in./s. At what rate is the area A of the square increasing when each edge is 9 in.?

Answer: 54 in.^{2}/s

18) A cubicle block of ice is melting in such a way that each edge decreases steadily by 3 in. every hour. At what rate is its volume decreasing when each edge is 9 in. long?

Answer: 729 in.^{3}/hr

19) A spherical balloon has sprung a leak. When the radius is 1 cm, the radius is changing at a rate of $-\dfrac{1}{4}$ cm/s. Find the rate of change of the volume at this time.

Answer: $\dfrac{dV}{dt} = -\pi$ cm/s

3.4 Derivatives of Algebraic Functions
Solve the problem.

1) Find $f'(x)$ given $f(x) =$ (a) $x^{5/3}$, (b) $4x^{-1/2}$, (c) $(x^2 + 1)^{1/2}$, (d) $(x^{3/2} + 1)^{2/3}$.

Answer: (a) $\dfrac{5x^{2/3}}{3}$, (b) $-\dfrac{2}{x^{3/2}}$, (c) $\dfrac{x}{\sqrt{x^2 + 1}}$, (d) $\dfrac{\sqrt{x}}{(x^{3/2} + 1)^{1/3}}$

2) Differentiate the function $f(x) = 3x^2 - \sqrt{x} + \dfrac{7}{\sqrt{x}}$.

Answer: $6x - \dfrac{1}{2}x^{-1/2} - \dfrac{7}{2}x^{-3/2} = 6x - \dfrac{1}{2\sqrt{x}} - \dfrac{7}{2x^{3/2}}$

3) Differentiate the function $h(z) = \dfrac{1}{\sqrt[3]{5+3z}}$.

Answer: $-(5+3z)^{-4/3} = -\dfrac{1}{(3z+5)^{4/3}}$

4) Differentiate the function $g(x) = (4x+3)^{5/3}$.

Answer: $\dfrac{20}{3}(4x+3)^{2/3}$

5) Differentiate the function $f(y) = (5-2x^3)^{-3/2}$

Answer: $9x^2(5-2x^3)^{-5/2}$

6) Differentiate the function $f(x) = \sqrt{4x^3+7}$.

Answer: $6x^2(4x^3+7)^{-1/2} = \dfrac{6x^2}{\sqrt{4x^3+7}}$

7) Differentiate the function $h(x) = \dfrac{x}{\sqrt{1+x^5}}$.

Answer: $\dfrac{2-3x^5}{2(1+x^5)^{3/2}}$

8) Differentiate the function $f(x) = x\sqrt{1-x^3}$.

Answer: $\dfrac{2-5x^3}{2\sqrt{1-x^3}}$

9) Differentiate the function $f(t) = \sqrt{\dfrac{t^2-1}{t^2+1}}$.

Answer: $2t(t^2-1)^{-1/2}(t^2+1)^{-3/2} = \dfrac{2t}{\sqrt{t^2-1}(t^2+1)^{3/2}}$

10) Differentiate the function $f(v) = \dfrac{\sqrt{3v-1}}{v}$.

Answer: $-\dfrac{3v-2}{2v^2\sqrt{3v-1}}$

11) Differentiate the function $f(x) = \sqrt[5]{1-x^2}$.

Answer: $-\dfrac{2x}{5}(1-x^2)^{-4/5} = -\dfrac{2x}{5(1-x^2)^{4/5}}$

12) Differentiate the function $f(x) = (1+x)^{3/2}(2+x)^{2/3}$

Answer: $\dfrac{(1+x)^{1/2}(22+13x)}{6(2+x)^{1/3}} = \dfrac{\sqrt{1+x}\,(22+13x)}{6(2+x)^{1/3}}$

13) Differentiate the function $f(x) = \dfrac{(3x+1)^{1/3}}{(4x-1)^{1/2}}$.

Answer: $-\dfrac{2x+3}{(3x+1)^{2/3}(4x-1)^{3/2}}$

14) Find all points on the graph of the curve defined by $y = x^{3/2}$ where the tangent line is either horizontal or vertical.

Answer: horizontal tangent at $(0, 0)$; no vertical tangents

15) Find all points on the graph of the curve defined by $y = x^{4/5}$ where the tangent line is either horizontal or vertical.

Answer: no horizontal tangents; vertical tangent at $(0, 0)$

16) Find all points on the graph of the curve defined by $y = \dfrac{x}{\sqrt{1-3x^2}}$ where the tangent line is either horizontal or vertical.

Answer: no horizontal or vertical tangents

17) Find all points on the graph of the curve defined by $y = x\sqrt{9-x^2}$ where the tangent line is either horizontal or vertical.

Answer: horizontal tangents at $(\dfrac{3}{\sqrt{2}}, \dfrac{9}{2})$ and $(-\dfrac{3}{\sqrt{2}}, -\dfrac{9}{2})$; vertical tangents at $(3, 0)$ and $(-3, 0)$

18) Write the equation of the line tangent to the curve $y = 3\sqrt{x}$ at the point P where $x = 4$. Write the equation in the form $ax + by = c$.

Answer: $3x - 4y = -12$

19) Write the equation of the line tangent to the curve $y = 5\sqrt{1+x}$ at the point P where $x = 3$. Write the equation in the form $ax + by = c$.

Answer: $5x - 4y = -25$

20) Match the graph of the function $y = x^{1/5}$ with the graph of its derivative.

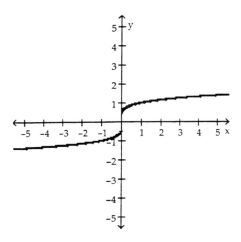

A)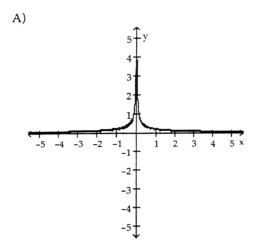

B)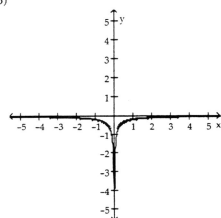

C)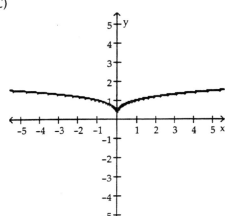

D)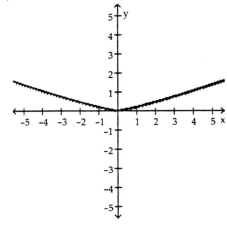

Answer: A

21) Match the graph of the function $y = x\sqrt{3-x}$ with the graph of its derivative.

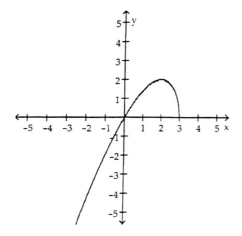

A)
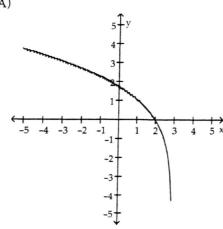

B)
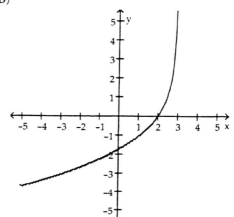

C)
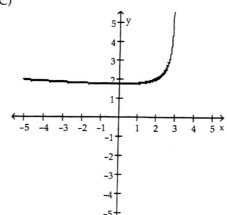

D)
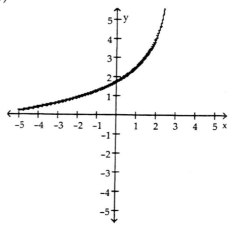

Answer: A

22) Match the graph of the function $y = x\sqrt{9-x^2}$ with the graph of its derivative.

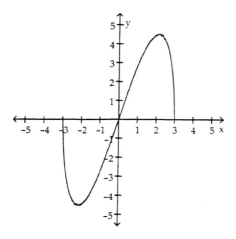

A)
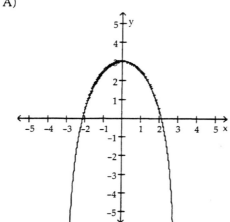

B)
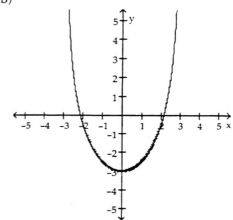

C)
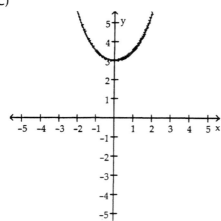

D)
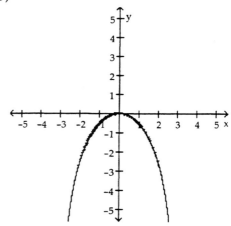

Answer: A

23) Find the two points on the circle $x^2 + y^2 = 1$ at which the slope of the tangent line is -3.

Answer: $\left(\dfrac{3}{\sqrt{10}}, \dfrac{1}{\sqrt{10}}\right), \left(-\dfrac{3}{\sqrt{10}}, -\dfrac{1}{\sqrt{10}}\right)$

24) Find two distinct lines through the point $P(0, \frac{1}{4})$ that are normal to the curve $y = x^{4/3}$.

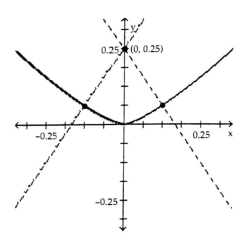

Answer: $6x - 4y = -1$, $6x + 4y = 1$

3.5 Maxima and Minima of Functions on Closed Intervals
Solve the problem.

1) State whether the function $f(x) = 3 - x$ attains a maximum value or a minimum value (or both) on the interval $[-2, 2)$.

   Answer: maximum is 5; no minimum

2) State whether the function $f(x) = |x| + 3$ attains a maximum value or a minimum value (or both) on the interval $(-2, 2)$.

   Answer: no maximum; minimum is 3

3) State whether the function $f(x) = 5 - x^3$ attains a maximum value or a minimum value (or both) on the interval $[-1, 2]$.

   Answer: maximum is 6; minimum is -3

4) State whether the function $f(x) = 3 - x^2$ attains a maximum value or a minimum value (or both) on the interval $[-2, 2)$.

   Answer: maximum is 3; minimum is -1

5) State whether the function $f(x) = \dfrac{9}{x^2 + 1}$ attains a maximum value or a minimum value (or both) on the interval $(-\infty, \infty)$.

   Answer: maximum is 9; no minimum

6) Find the maximum and minimum values attained by the function $f(x) = -5x + 3$ on the interval $[-2, 4]$.

   Answer: max = 13; min = -17

7) Find the maximum and minimum values attained by the function $f(x) = 3x^2 - 12x$ on the interval $[0, 5]$.

   Answer: max = 15; min = -12

8) Find the maximum and minimum values attained by the function $f(x) = (3x - 2)^2$ on the interval $[-1, 4]$.

   Answer: max = 100; min = 0

9) Find the maximum and minimum values attained by the function $f(x) = x^3 + x - 1$ on the interval $[0, 1]$.

   Answer: max = 1; min = -1

10) Find the maximum and minimum values attained by the function $h(x) = 2x^3 - 3x^2 - 12x + 4$ on the closed interval $[0, 3]$.

    Answer: max = 4; min = -16

11) Find the maximum and minimum values attained by the function $h(x) = x + \dfrac{5}{x}$ on the interval $[1, 5]$.

    Answer: max = 6; min = $\dfrac{10}{\sqrt{5}}$

12) Find the maximum and minimum values attained by the function $g(x) = x^3 - 3x$ on the interval $[-2, 4]$.

    Answer: min = -2; max = 52

13) Find the maximum and minimum values attained by the function $f(x) = -3x^5 + 20x^3$ on the interval $[-3, 3]$.

    Answer: max = 189; min = -189

14) Find the maximum and minimum values attained by the function $f(x) = 3 - |9 - 5x|$ on the interval $[0, 4]$.

    Answer: max = 3; min = -8

15) Find the maximum and minimum values attained by the function $h(x) = \dfrac{x - 1}{x + 1}$ on the interval $[0, 2]$.

    Answer: min = -1; max = $\dfrac{1}{3}$

16) Find the maximum and minimum values attained by the function $f(x) = \dfrac{(x - \frac{3}{4})}{x^2 + 1}$ on the interval $[-1, 3]$.

    Answer: max = $\dfrac{1}{4}$; min = -1

17) Find the maximum and minimum values attained by the function $g(x) = x^{1/3}$ on the interval $[-1, 8]$.

    Answer: min = -1; max = 2

18) Find the maximum and minimum values attained by the function $f(x) = x\sqrt{9 - x^2}$ on the interval $[0, 3]$.

    Answer: max = $\dfrac{9}{2}$; min = 0

19) Find good approximations for the maximum and minimum values of the function $f(x) = x^3 + 5x^2 - 7x + 8$ on the interval $[-5, 2]$ by zooming in on the zeros of the derivative.

    Answer: max = 52.03; min = 5.82

Calculus 59

20) Find good approximations for the maximum and minimum values of the function $f(x) = x^4 - 5x^3 + 13x - 7$ on the interval [0, 4] by zooming in on the zeros of the derivative.

Answer: max = 2.11; min = -25.82

21) Find good approximations to the maximum and minimum values of the function
$f(x) = x^5 - 2x^4 - 9x^3 + 13x^2 + 7x + 5$ on the interval [0, 3] by zooming in on the zeros of the derivative.

Answer: max = 15.14; min = -21.88

3.6 Applied Optimization Problems
Solve the problem.

1) Find the maximum possible area of a rectangle with perimeter 180 meters.

Answer: $A = 2025$ m^2

2) A rectangular corral is to be constructed with an internal divider parallel to two opposite sides, and 1200 meters of fencing will be used to make the corral (including the divider). What is the maximum possible area that such a corral can have?

Answer: 60,000 m^2

3) The sum of the squares of two nonnegative real numbers x and y is 18. What is the minimum possible value of $x + y$?

Answer: $\sqrt{18}$

4) What is the greatest amount by which a number in the interval [0, 1] can exceed its cube?

Answer: $\dfrac{2\sqrt{3}}{9}$

5) A farmer has 600 yards of fencing with which to build a rectangular corral. He will use a long straight wall to form one side of the corral (to save fencing). What is the maximum possible area that can be enclosed in this way?

Answer: 45,000 yd^2

6) Two nonnegative numbers have sum 1. What is the maximum possible value of the sum of their squares?

Answer: 1

7) The stiffness of a rectangular beam is proportional to the product of its width and the cube of the height of its cross section. What shape beam should be cut from a circular log of radius R to maximize the stiffness of the beam?

Answer: $w = R$, $h = \sqrt{3}R$

8) Find the shape of the right circular cylinder of maximal surface area (including the top and the bottom) inscribed in a sphere of radius R.

Answer: $r = \sqrt{\dfrac{1}{2} + \dfrac{1}{2\sqrt{5}}} R$, $h = 2R \left(\dfrac{5 - \sqrt{5}}{10} \right)^{1/2}$

(Must satisfy $r^2 + \left(\dfrac{1}{2}h\right)^2 = R^2$)

9) Two cubes have total volume 250 cm³. What is the maximum possible surface area they can have? The minimum?

   Answer: min = $150 \sqrt[3]{4}$ cm², max = 300 cm²

10) A rectangle has each diagonal of length 5. What is the maximum possible perimeter of such a rectangle?

    Answer: max = $10\sqrt{2}$

11) The product of two positive integers is 1600. What is the minimum possible value of their sum?

    Answer: $S = 80$

12) A rectangle has its base on the x-axis and its upper two vertices on the graph of $y = 12 - x^2$. What is the maximum possible area of such a rectangle?

    Answer: 32

13) The sum of two nonnegative numbers x and y is 8. Find the maximum possible value of the expression $x^2 + y^3$.

    Answer: max = 512

14) A mass of clay of volume 432 in.³ is formed into two cubes. What is the minimum possible total surface area of the two cubes? What is the maximum?

    Answer: max = 432 in.², min = $216 \sqrt[3]{4}$ in.²

15) A rancher has 1700 meters of fencing with which to build two widely separated corrals; one is to be square, and the other is to be twice as long as it is wide. What is the maximum possible total area that the rancher can thereby enclose?

    Answer: 180,625 m² (using all the fence for the square corral)

16) Find the maximum possible volume of a right circular cylinder inscribed in a sphere of radius R.

    Answer: $\frac{4\sqrt{3}}{9} \pi R^3$

17) A farmer has 480 meters of fencing. He wishes to enclose a rectangular plot of land and to divide the plot into three equal rectangles with two parallel lengths of fence down the middle. What dimensions will maximize the enclosed area? Be sure to verify that you have found the *maximum* enclosed area.

    Answer: $A = 7,200$, width = 60, length = 120

18) The sum of two nonnegative numbers is 10. Find the minimum possible value of the sum of their cubes.

    Answer: $S = 250$

19) The sum of two nonnegative numbers is 48. What is the smallest possible value of the sum of their squares?

    Answer: $S = 1,152$

20) A rectangle has a line of fixed length L reaching from one vertex to the midpoint of one of the far sides. What is the maximum possible area of such a rectangle?

    Answer: L^2

Calculus 61

21) Write an equation of the straight line through the point (7, 1) that cuts off the least area from the first quadrant. Be sure to verify that your area is *minimal*.

    Answer: $x + 7y = 14$

22) A circle is dropped into the graph of the parabola $y = x^2$. How small can the radius of the circle be and yet allow the circle to touch the parabola at two different points?

    Answer: $r > \dfrac{1}{2}$

23) Find the point on the graph of $y = \sqrt{x}$ that is closest to the point (3, 0). Be sure to verify that it is indeed closest.

    Answer: $\left(\dfrac{5}{2}, \sqrt{\dfrac{5}{2}}\right)$

24) A rectangular box is to have a base three times as long as it is wide. The total surface area of the box is to be 20 ft². Find the maximum possible volume of such a box.

    Answer: $\dfrac{5\sqrt{10}}{3}$ ft³

25) A circular sector is bounded by two radii of the circle (each of length r) and by a circular arc of length s. The area of such a figure is known to be $\dfrac{1}{2}rs$. Find the value of r that will produce such a sector of maximum possible area, given that its perimeter must be 100.

    Answer: $r = 25$

26) Find two nonnegative numbers whose sum is 4 such that the sum of the square of one and the cube of the other is (a) maximal; (b) minimal.

    Answer: (a) 0 and 4, (b) $\dfrac{8}{3}$ and $\dfrac{4}{3}$

27) A right circular cone has a slant height of 10 ft. Find the maximum possible volume of such a cone. Verify that your answer is maximal.

    Answer: $\dfrac{2000\pi}{9\sqrt{3}}$

28) A rectangular poster is to contain 8250 cm² of print in the shape of a smaller rectangle; the margins at top and bottom must each be 22 cm, and those at the sides must each be 15 cm. What are the dimensions of such a poster having the least possible total area?

    Answer: 105 cm wide × 154 cm high

29) A wastebasket is to have as its base an equilateral triangle, its sides are to be vertical, and its volume is to be 8 ft³. What is the minimum possible surface area of such a wastebasket?

    Answer: $12\sqrt{3}$ ft²

30) Write the number 33 as the sum of two nonnegative real numbers such that the product of one number with the square of the other is maximal.

    Answer: 11 + 22

Calculus 62

31) Find the coordinates of the point or points on the curve $2y^2 = 5x + 5$ which is (are) closest to the origin (0, 0).

   Answer: (−1, 0)

32) A railroad will operate a special a special excursion train if at least 200 people subscribe. The fare will be $8 per person if 200 people subscribe, but will decrease 1 cent for each additional person who subscribes. What number of passengers will bring the railroad maximum revenue?

   Answer: 500

3.7 Derivatives of Trigonometric Functions
Solve the problem.

1) Differentiate the function $f(x) = x \sin x$.

   Answer: $\sin x + x \cos x$

2) Differentiate the function $f(x) = \cos^2 x \sin^3 x$

   Answer: $-2 \cos x \sin^4 x + 3 \sin^2 x \cos^3 x$

3) Differentiate the function given $g(x) =$ (a) $x^3 \sin x$, (b) $\sin^2 x$, (c) $\dfrac{1}{\sin x}$, (d) $(\sin x)(x \sin x)$.

   Answer: (a) $x^3 \cos x + 3x^2 \sin x$ (b) $2 \sin x \cos x$ (c) $-(\cot x \csc x)$
   (d) $2x \sin x \cos x + \sin^2 x$

4) Find dy/dx given $y =$
   (a) $\sin(x^4)$, (b) $(\sin x)^4$, (c) $x^3 \sin x^3$, (d) $(\sin x)(\sin x^2)$.

   Answer: (a) $4x^3 \cos(x^4)$, (b) $4 \cos x \sin^3 x$, (c) $3x^5 \cos(x^3) + 3x^2 \sin(x^3)$,
   (d) $2x \cos x^2 \sin x + \cos x \sin(x^2)$

5) Find dy/dx given $y = (\cos x - \sin x)^2$.

   Answer: $-2(\cos^2 x - \sin^2 x) = -2 \cos 2x$

6) Find dy/dx given $y =$
   (a) $\sin 7x$ (b) $\cos 5x^2$
   (c) $(\sin 7x)^3$ (d) $\sqrt{3 \cos 3x}$

   Answer: (a) $7 \cos 7x$ (b) $-10x \sin(5x^2)$
   (c) $21 \cos 7x \sin^2 7x$ (d) $-\dfrac{3^{3/2} \sin(3x)}{\sqrt{2 \cos 3x}}$

7) Find dy/dx given $y =$
   (a) $\sin 2\pi x$, (b) $(\sin x)^{100}$, (c) $\sin(x^{100})$, (d) $(x \sin x)^{4/3}$

   Answer: (a) $2\pi \cos(2\pi x)$ (b) $100 \cos x \sin^{99} x$ (c) $100 x^{99} \cos(x^{100})$
   (d) $\dfrac{4(x \sin x)^{1/3} (x \cos x + \sin x)}{3}$

8) Find dy/dx given $y =$
   (a) $(x \sin x^2)^7$, (b) $\sin x^7$, (c) $\sin^7 x$, (d) $(\sin 5x)^3$, (e) $x^2 \sin^3 4x$, (f) $\sin 7x$, (g) $\sin 7x \sin 8x$.

   Answer: (a) $14x^8 \cos(x^2) \sin^6(x^2) + 7x^6 \sin^7(x^2)$
   (b) $7x^6 \cos(x^7)$
   (c) $7 \cos x \sin^6 x$
   (d) $15 \cos(5x) \sin^2(5x)$
   (e) $12x^2 \cos(4x) \sin^2(4x) + 2x \sin^3(4x)$
   (f) $7 \cos x$
   (g) $8 \cos(8x) \sin(7x) + 7 \cos(7x) \sin(8x) = \dfrac{15 \sin(15x) - \sin x}{2}$

9) Find dy/dx given $y =$
   (a) $\sin^2 x \cos^2 x$ (b) $\dfrac{\sin 2x}{\cos 3x}$
   (c) $\sin x \cos 3x$ (d) $\cos(\cos x)$

   Answer: (a) $2 \cos^3 x \sin x - 2 \cos x \sin^3 x$ (b) $2 \cos x \sin 3x + 3 \sec 3x \sin 2x \tan 3x$
   (c) $\cos x \cos 3x - 3 \sin x \sin 3x$ (d) $\sin x \sin(\cos x)$

10) Find dy/dx given $y =$
   (a) $x^2 \sin x^3$ (b) $\sin^7 x \cos x$
   (c) $\cos^4 x^4$ (d) $\sin(x^3 + x^2 + x + 1)$

   Answer: (a) $3x^4 \cos x^3 + 2x \sin x^3$ (b) $7 \cos^2 x \sin^6 x - \sin^8 x$
   (c) $-16x^3 \cos^3 x^4 \sin x^4$ (d) $(3x^2 + 2x + 1) \cos(x^3 + x^2 + x + 1)$

11) Find dy/dx given $y = \sqrt{x}(x - \sin x)^3$.

   Answer: $\dfrac{1}{2}x^{-1/2}(x - \sin x)^3 + 3x^{1/2}(x - \sin x)^2(1 - \cos x) = \dfrac{(x - \sin x)^2(7x - \sin x - 6x \cos x)}{2\sqrt{x}}$

12) Find the derivatives of the functions
   (a) $F(x) = (1 - 2 \sin x)^4$, (b) $G(t) = (\sin t^7)^{19}$, (c) $H(z) = (\sin z \sin 3z)^9$.

   Answer: (a) $-8 \cos x(1 - 2\sin x)^3$, (b) $133 t^6 \cos(t^7) \sin^{18}(t^7)$,
   (c) $27 \cos(3z) \sin^9 z \sin^8(3z) + 9 \cos z \sin^8 z \sin^9(3z)$

13) Find dy/dx given $y =$
   (a) $\sqrt{\sin x}$, (b) $\sin \sqrt{x}$, (c) $x^4(1 + \sin x)^{3/2}$, (d) $\sin(x + x^{1/2} + 1)$.

   Answer: (a) $\dfrac{\cos x}{2\sqrt{\sin x}}$, (b) $\dfrac{\cos \sqrt{x}}{2\sqrt{x}}$, (c) $\dfrac{3x^4 \cos x \sqrt{\sin x + 1}}{2} + x^3(1 + \sin x)^{3/2}$,
   (d) $\left[1 + \dfrac{1}{2\sqrt{x}}\right] \cos(1 + \sqrt{x} + x)$

14) Find dx/dt given $x = \tan t^6$.

   Answer: $6t^5 \sec^2 t^6$

15) Find dx/dt given $x = (\sec 3t)^6$.

   Answer: $18 \sec^6 3t \tan 3t$

Calculus 64

16) Find dy/dx given $y =$
(a) $\sec x^2$, (b) $\tan^3 x$, (c) $\sec \dfrac{1}{x}$, (d) $\sec x \tan x$.

Answer: (a) $2x \sec(x^2) \tan(x^2)$, (b) $3 \sec^2 x \tan^2 x$, (c) $-\dfrac{\sec(1/x) \tan(1/x)}{x^2}$,

(d) $\sec^3 x + \sec x \tan^2 x$

17) Find dx/dt given $x = \csc\left(\dfrac{1}{t^3}\right)$.

Answer: $\dfrac{3 \csc\left(\dfrac{1}{t^3}\right) \cot\left(\dfrac{1}{t^3}\right)}{t^4}$

18) Find dx/dt given $x = \cot\left(\dfrac{1}{\sqrt{2t}}\right)$.

Answer: $\dfrac{\csc^2\left(\dfrac{1}{\sqrt{2t}}\right)}{(2t)^{3/2}}$

19) Find dx/dt given $x = \dfrac{\sec 3t}{\tan 5t}$.

Answer: $\dfrac{3 \sec 3t \tan 3t \tan 5t - 5 \sec^2 5t \sec 3t}{\tan^2 5t}$

20) Find dx/dt given $x = \sec(\cos 3t)$.

Answer: $-3[\sec(\cos 3t) \tan(\cos 3t)] \sin 3t$

21) Find dy/dx given $y =$
(a) $\sec(x^2 + 5x)$, (b) $\sec^5(x^2 + 5x)$,
(c) $\sec 3x \tan 5x$, (d) $\tan^3 x^3$.

Answer: (a) $(2x+5) \sec(x^2+5x) \tan(x^2+5x)$, (b) $5(2x+5) \sec^5(x^2+5x) \tan(x^2+5x)$,
(c) $5 \sec 3x \sec^2 5x \; 3 \sec 3x \tan 3x \tan 5x$ (d) $9x^2 \sec^2 x^3 \tan^3 x^3$

22) Find dy/dx given $y =$ (a) $\sec \sqrt{x}$, (b) $\sqrt{\sec x}$, (c) $\sec^7 x$,
(d) $\sec x^7$, (e) $\tan^8 x$, (f) $\tan x^8$.

Answer: (a) $\dfrac{\sec \sqrt{x} \tan \sqrt{x}}{2\sqrt{x}}$, (b) $\dfrac{(\sec x)^{3/2} \sin x}{2}$, (c) $7 \sec^7 x \tan x$, (d) $7x^6 \sec x^7 \tan x^7$,
(e) $8 \sec^2 x \tan^7 x$, (f) $8x^7 \sec^2(x^2)$

23) Find dy/dx given $y =$
(a) $\sin x \cos x \sec x$, (b) $\sec^2 x - \tan^2 x$,
(c) $\dfrac{\sin x + \cos x}{\sec x + \tan x}$, (d) $\sec^4 \sqrt{x}$.

Answer: (a) $\cos x$ (b) 0,

(c) $\dfrac{\cos x - \sin x}{\sec x + \tan x} - \dfrac{(\cos x + \sin x)(\sec^2 x + \sec x \tan x)}{(\sec x \tan x)^2} =$

$\dfrac{1 + \cos 2x - \sin 2x - 2(\sin x + \cos x)}{2 + \sin 2x}$,

(d) $\dfrac{2\sec^4 \sqrt{x} \tan \sqrt{x}}{\sqrt{x}}$

24) Find dx/dt given $x = \sqrt{1 + \cot 3t}$.

Answer: $-\dfrac{3 \csc^2 3t}{2\sqrt{1 + \cot 3t}}$

25) Write an equation of the line that is tangent to the curve $y = x \sin x$ at the point P with x-coordinate $\dfrac{\pi}{2}$.

Answer: $y = x$

26) Find all points on the curve $y = \sin 2x$ where the tangent line is horizontal.

Answer: At every odd integral multiple of $\dfrac{\pi}{4}$.

27) A rocket is launched vertically and is tracked by a radar station located on the ground 6 mi from the launch pad. Suppose that the elevation angle θ of the line of sight to the rocket is increasing at 5° per second when $\theta = 60°$. What is the velocity of the rocket at this instant?

Answer: $\dfrac{2\pi}{3}$ mi/s $\approx$ 7540 mi/h

28) An isosceles triangle is inscribed in a circle of radius 5. Determine the maximum possible area of the triangle.

Answer: $\dfrac{75\sqrt{3}}{4}$

29) Find $f'(x)$ given $f(x) =$
(a) $\csc^3 x$, (b) $\cot 7x^2$, (c) $\csc 3x \cot 5x$, (d) $\cot^3 x^4$

Answer: (a) $-3\cot x \csc^3 x$, (b) $-14x \csc^2 7x^2$,
(c) $-3\cot 3x \cot 5x \csc 3x - 5\csc 3x \csc^2 5x$, (d) $-12x^3 \cot^2 x^4 \csc^3 x^4$

30) Find $g'(x)$ given $g(x) =$

(a) $x^6 \cot 7x$, (b) $\sqrt{\csc x \cot x}$, (c) $\dfrac{1 + \cot x}{1 + \csc x}$, (d) $\csc^8 x \cot^4 x$

Answer: (a) $6x^5 \cot 7x - 7x^6 \csc^2 7x$, (b) $\dfrac{-(\cot^2 x \csc x) - \csc^3 x}{2\sqrt{\cot x \csc x}} = \dfrac{-\csc^3 x(3 + 2\cos 2x)}{4\sqrt{\cot x \csc x}}$,

(c) $\dfrac{\cot x(1 + \cot x) \csc x}{(\csc x + 1)^2} - \dfrac{\csc^2 x}{\csc x + 1} = \dfrac{-2 \sin\left(\dfrac{x}{2}\right)}{\left(\sin\left(\dfrac{x}{2}\right) + \cos\left(\dfrac{x}{2}\right)\right)^3}$,

(d) $-8 \cot^5 x \csc^8 x - 4\cot^3 x \csc^{10} x$

3.8 Successive Approximations and Newton's Method
Solve the problem.

1) Use Newton's method to find the solution of the equation $x^2 - 7 = 0$ in the interval $[2, 3]$ accurate to four decimal places.

   Answer: 2.6458

2) Use Newton's method to find the solution of the equation $x^5 - 50 = 0$ in the interval $[2, 3]$ accurate to four decimal places.

   Answer: 2.1867

3) Use Newton's method to find the solution of the equation $x^2 + 4x - 1 = 0$ in the interval $[0, 1]$ accurate to four decimal places.

   Answer: 0.2361

4) Use Newton's method to find the solution of the equation $x^3 + 3x + 5 = 0$ in the interval $[1, 2]$ accurate to four decimal places.

   Answer: 1.1542

5) Use Newton's method to find the solution of the equation $x^3 - \sin x = 0$ in the interval $[0.5, 1]$ accurate to four decimal places. [*Remember*: Set your calculator in *radian* mode.]

   Answer: 0.9286

6) Use Newton's method to find the solution of the equation $3 + x \sin x = 0$ in the interval $[3, 4]$ accurate to four decimal places. [*Remember*: Set your calculator in *radian* mode.]

   Answer: 3.9919

7) Use Newton's method to find the solution of the equation $x^5 + 2x = 1$ in the interval $[0, 1]$ accurate to four decimal places.

   Answer: 0.4864

8) Use Newton's method to find the value of x for which $\cos x = x$ with four digits correct to the right of the decimal point. [*Remember*: Set your calculator in *radian* mode.]

   Answer: 0.7391

9) Use Newton's method to find all real roots of $\sin x = \frac{1}{3}x$ with four digits correct to the right of the decimal point.

Answer: −2.2789, 0, 2.2789

10) Use Newton's method to approximate $\sqrt[3]{122}$ with four place accuracy.

Answer: 4.9597

11) The equation $f(x) = x^3 - 3x + 1$ has three distinct real roots. Approximate their locations by evaluating f at −2, −1, 0, 1, and 2. Then use Newton's method to approximate each of the three roots to four-place accuracy.

Answer: −1.8794, 0.3473, 1.5321

Ch. 4 Additional Applications of the Derivative
4.1 Implicit Differentiation and Related Rates
Solve the problem.

1) Given the curve $x^2 + y^2 = 16$, find dy/dx by implicit differentiation.

   Answer: $\dfrac{dy}{dx} = -\dfrac{x}{y}$

2) Given $x^{2/3} + y^{2/3} = 1$, find dy/dx by implicit differentiation.

   Answer: $\dfrac{dy}{dx} = -\dfrac{y^{1/3}}{x^{1/3}}$

3) Given $xy + y^2 + x^3 = 7$, find dy/dx by implicit differentiation.

   Answer: $\dfrac{dy}{dx} = -\dfrac{3x^2 + y}{x + 2y}$

4) Given $x^2 y^2 = x^2 + y^4$, find dy/dx by implicit differentiation and simplify.

   Answer: $\dfrac{x - xy^2}{x^2 y - 2y^3}$

5) Given $y = \sin(x^2 + y)$, find dy/dx by implicit differentiation.

   Answer: $\dfrac{dy}{dx} = \dfrac{2x \cos(x^2 + y)}{1 - \cos(x^2 + y)}$

6) Use implicit differentiation to find an equation of the line tangent to the curve $x^2 + y^2 = 169$ at the point $(5, -12)$.

   Answer: $5x - 12y = 169$

7) Use implicit differentiation to find an equation of the line tangent to the curve $xy = -12$ at the point $(3, -4)$.

   Answer: $4x - 3y = 24$

8) Use implicit differentiation to find an equation of the line tangent to the curve $3x^2 y = x + 2$ at the point $(1, 1)$.

   Answer: $5x + 3y = 8$

9) Use implicit differentiation to find an equation of the line tangent to the curve $x^3 + 2xy + y^3 = 13$ at the point $(1, 2)$.

   Answer: $x + 2y = 5$

10) Use implicit differentiation to find an equation of the line tangent to the curve $\dfrac{1}{2x + 3} + \dfrac{1}{2y - 3} = -2$ at the point $(-2, 1)$.

    Answer: $x + y = -1$

11) Find all points on the graph of $x^2 y + 2 = xy^2$ at which the tangent line is horizontal.

    Answer: $(1, 2)$

Calculus 69

12) Sand falling from a hopper at 10π ft^3/s forms a conical sandpile whose radius is always equal to its height. How fast is the radius increasing when the radius is 5 ft?

Answer: $\frac{2}{5}$ ft/s

13) A circular oil slick spreads on the surface of the ocean, the result of a spill of 10 ft^3 of oil. When its radius is 50 ft, its radius is increasing at the rate of 2 ft/s. How fast is the thickness of the slick changing then?

Answer: It is decreasing at $\frac{1}{3125\pi}$ ft/s

14) Each edge of a square is increasing at the rate of 10 in./s. How fast is its area increasing when each edge has length 50 in.?

Answer: 1,000 in^2/s

15) The sides of an equilateral triangle are lengthening at the rate of 3 cm/min. How fast is the area of the triangle increasing when its sides are each 8 cm long?

Answer: $12\sqrt{3}$ in^2/min

16) The base of a rectangle is growing larger at the rate of 5 cm/min while its altitude is growing smaller at the rate of 3 cm/min. At what rate is the area of the rectangle changing when its base is 40 cm long and its altitude is 90 cm?

Answer: 330 in^2/min

17) A rocket is launched vertically and is tracked by a radar station located on the ground 4 km from the launch site. What is the speed of the rocket at the instant when its distance from the radar station is 5 km if this distance is increasing at the rate of 3600 km/h then?

Answer: 6,000 km/h

18) Two aircraft approach Foley Field at the same constant altitude. The first aircraft is moving south at 250 km/h while the second is moving west at 600 km/h. At what rate is the distance between then changing when the first aircraft is 60 km from the field and the second is 25 km from the field?

Answer: $\frac{600}{13}$ mi/h

19) A telephone pole is 15 ft away from a street light; the latter is 20 ft above the ground. A squirrel runs up the telephone pole at 8 ft/s. How fast is the squirrel's shadow traveling along the (level) ground when the squirrel is 18 ft above the ground?

Answer: 600 ft/s

20) A ladder 13 ft long is sliding down a vertical wall while its base slides horizontally along the ground. The top of the ladder is moving at 12 ft/s when it is 5 ft above the ground. How fast is the base of the ladder moving then?

Answer: 5 ft/s

4.2 Increments, Differentials, and Linear Approximation
Solve the problem.

1) If $y = 4x^3 - \dfrac{3}{x^2}$, write dy in terms of x and dx.

   Answer: $dy = (12x^2 + 6x^{-3})dx$

2) Write the linear approximation L(x) to the function $f(x) = (1 + x)^3$ near the point a = 0.

   Answer: $L(x) = 1 + 3x$

3) Use a linear approximation L(x) to an appropriate function f(x), with an appropriate value of a, to estimate the number $\sqrt{82}$. Round to four decimal places.

   Answer: $\dfrac{163}{18} \approx 9.0556$

4) Use a linear approximation L(x) to an appropriate function f(x), with an appropriate value of a, to estimate the number $\sqrt[5]{31}$. Round to four decimal places.

   Answer: $\dfrac{159}{80} = 1.9875$

5) Use a linear approximation L(x) to an appropriate function f(x), with an appropriate value of a, to estimate the number $\sqrt[3]{999}$. Round to four decimal places.

   Answer: $\dfrac{2999}{300} \approx 9.9967$

6) Use a linear approximation L(x) to an appropriate function f(x), with an appropriate value of a, to estimate the number sin 58°. Round to four decimal places.

   Answer: $\dfrac{\sqrt{3} - \frac{1}{90}\pi}{2} \approx 0.8486$

7) Compute the differential of each side of the equation $x^4 + y^4 = 4xy$, regarding x and y as dependent variables. Then solve for dy/dx.

   Answer: $dy/dx = \dfrac{y - x^3}{y^3 - x}$

8) Use a linear approximation to estimate the change in the area of a circle, if its radius is decreased from 10 in. to 9.8 in.

   Answer: -4π in.$^2 \approx -12.5664$ in.^{2}

9) The radius of a spherical ball is measured as 10 in., with a maximum error of $\dfrac{1}{16}$ in. What is the maximum resulting error in its calculated surface area?

   Answer: $\pm 5\pi$ in.$^2 \approx \pm 15.7080$ in.^{2}

10) For the function $f(x) = 3x^2$ and the point a = 1, determine graphically an open interval I, centered at a, so that f(x) and its linear approximation L(x) differ by less than the the value $\varepsilon = 0.1$ at each point in I.

    Answer: (0.82, 1.18)

11) For the function f(x) = cos x and the point $a = \frac{\pi}{3}$, determine graphically an open interval I, centered at a, so that f(x) and its linear approximation L(x) differ by less than the the value ε = 0.02 at each point in I.

Answer: (0.79, 1.3)

4.3 Increasing and Decreasing Functions and the Mean Value Theorem
Solve the problem.

1) For the functions below, determine the open intervals on the x-axis on which each function is increasing and decreasing. Then use this information to match the function to its graph.

   1) $f(x) = 3 - x^2$ 2) $f(x) = x^2 - 4x - 3$ 3) $f(x) = 2x^3 - 3x$ 4) $f(x) = 2x^2 - 3x - \frac{1}{3}x^3$

A)
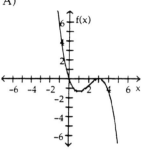

B)
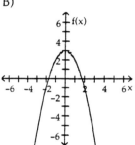

C)
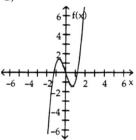

D)
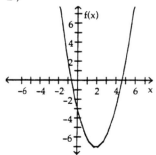

Answer: 1B, 2D, 3C, 4A

Calculus 73

2) Given $f'(x) = 3x$ and $f(0) = -2$, find $f(x)$.

   Answer: $f(x) = \frac{3}{2}x^3 - 2$

3) Given $f'(x) = \frac{1}{x^2}$ and $f(1) = 3$, find $f(x)$.

   Answer: $f(x) = -\frac{1}{x} + 4$

4) Find the open intervals on the x-axis on which the function $f(x) = 2x - 7$ is increasing and those on which it is decreasing.

   Answer: increasing on $\mathcal{R}$

5) Find the open intervals on the x-axis on which the function $f(x) = 3x^2 + 12x + 5$ is increasing and those on which it is decreasing.

   Answer: increasing on $(-2, \infty)$, decreasing on $(-\infty, -2)$

6) Find the open intervals on the x-axis on which the function $f(x) = -x\sqrt{x^2 + 4}$ is increasing and those on which it is decreasing.

   Answer: decreasing on $\mathcal{R}$

7) Find the open intervals on the x-axis on which the function $f(x) = x^4 + 4x^3$ is increasing and those on which it is decreasing.

   Answer: increasing on $(-3, 0) \cup (0, \infty)$, decreasing on $(-\infty, -3)$

8) Find the open intervals on the x-axis on which the function $f(x) = x^3 - 27x$ is increasing and those on which it is decreasing.

   Answer: increasing on $(-\infty, -3) \cup (3, \infty)$, decreasing on $(-3, 3)$

9) Find the open intervals on the x-axis on which the function $f(x) = x^{2/3}$ is increasing and those on which it is decreasing.

   Answer: increasing on $(0, \infty)$, decreasing on $(-\infty, 0)$

10) Find the open intervals on the x-axis on which the function $f(x) = \frac{x^2}{x - 1}$ is increasing and those on which it is decreasing.

    Answer: increasing on $(-\infty, 0) \cup (2, \infty)$, decreasing on $(0, 1) \cup (1, 2)$

11) Find the open intervals on the x-axis on which the function $f(x) = \frac{x^2 - 1}{x^2 + 1}$ is increasing and those on which it is decreasing.

    Answer: increasing on $(0, \infty)$, decreasing on $(-\infty, 0)$

12) Show that the function $f(x) = x^2 - 5x$ satisfies the hypotheses of Rolle's theorem on the interval $[0, 5]$, and find all numbers in $(0, 5)$ that satisfy the conclusion of that theorem.

    Answer: $f(0) = 0 = f(5)$; $f'(x) = 2x - 5$; $c = \frac{5}{2}$

13) Show that the function $f(x) = 3 - |x|$ does not satisfy the conclusion of Rolle's theorem on the interval $[-3, 3]$. Which of the hypotheses does it fail to satisfy?

Answer: $f'(0)$ does not exist.

14) Show that the function $f(x) = x^3$ satisfies the hypotheses of the mean value theorem on the interval $[0, 2]$. Find all numbers c in that interval that satisfy the conclusion of that theorem.

Answer: $c = \dfrac{2\sqrt{3}}{3}$

15) Show that the function $f(x) = \sqrt{x}$ satisfies the hypotheses of the mean value theorem on the interval $[0, 4]$. Find all numbers c in that interval that satisfy the conclusion of that theorem.

Answer: $c = 1$

16) Show that the function $f(x) = x^4$ satisfies the hypotheses of the mean value theorem on the interval $[0, 2]$. Find all numbers c in that interval that satisfy the conclusion of that theorem.

Answer: $c = \sqrt[3]{2}$

17) Show that the function $f(x) = x^{1/3}$ satisfies the hypotheses of the mean value theorem on the interval $[0, 27]$. Find all numbers c in that interval that satisfy the conclusion of that theorem.

Answer: $c = 3\sqrt{3}$

18) Show that the function $f(x) = x^3 + x + 1$ satisfies the hypotheses of the mean value theorem on the interval $[-1, 2]$. Find all numbers c in that interval that satisfy the conclusion of that theorem.

Answer: $c = \pm 1$

19) Show that the function $f(x) = x - \dfrac{1}{x}$ satisfies the hypotheses of the mean value theorem on the interval $[1, 3]$. Find all numbers c in that interval that satisfy the conclusion of that theorem.

Answer: $c = \sqrt{3}$

20) Show that the function $f(x) = x^3 + x - 4$ satisfies the hypotheses of the mean value theorem on the interval $[-2, 3]$. Find all numbers c in that interval that satisfy the conclusion of that theorem.

Answer: $c = \pm \dfrac{\sqrt{21}}{3}$

21) Show that the function $f(x) = \dfrac{11}{5}x^5$ satisfies the hypotheses of the mean value theorem on the interval $[-1, 2]$. Find all numbers c in that interval that satisfy the conclusion of that theorem.

Answer: $c = \sqrt[4]{\dfrac{11}{5}}$

22) Show that the equation $2x^3 - 5x - 4 = 0$ has exactly one solution in the interval $[1, 2]$.

Answer: If $g(x) = 2x^3 - 5x - 4$, then $g(1) = -7$ and $g(2) = 2$. So g has at least one zero because $g(1) < 0 < g(2)$, and g is continuous. Since $g'(x) = 6x^2 - 5 > 0$ for all x in $[1, 2]$, g is an increasing function on $[1, 2]$. Hence, g can have at most one zero in $[1, 2]$.

23) A car is traveling along a stretch of highway where the speed limit is 65 mph. At 1:00 P.M., its odometer (measuring its distance traveled) reads 14,207 mi. At 1:22, it reads 14,232 mi. Prove that the driver violated the speed limit at some instant between 1:00 and 1:22 P.M.

Answer: The average speed of the car is given by $\dfrac{14{,}232 - 14{,}207}{\frac{22}{60} - 0} \approx 68.18$ mph. The mean value theorem states that the the car must have been traveling at this speed, which exceeds the 65 mph limit, at some instant between 1:00 and 1:22 P.M.

4.4 The First Derivative Test and Applications
Solve the problem.

1) Classify each of the critical points of the function $f(x) = x^2 + 6x - 1$ as global or local, maximum or minimum, or not an extremum.

   Answer: global minimum at $x = -3$

2) Classify each of the critical points of the function $f(x) = 10 + 45x - 3x^2 - x^3$ as global or local, maximum or minimum, or not an extremum.

   Answer: local minimum at $x = -5$; local maximum at $x = 3$

3) Classify each of the critical points of the function $f(x) = 4x^3 - 3x^4$ as global or local, maximum or minimum, or not an extremum.

   Answer: no local extremum at $x = 0$; local maximum at $x = 1$

4) Classify each of the critical points of the function $f(x) = x^2 + \dfrac{16}{x}$ as global or local, maximum or minimum, or not an extremum.

   Answer: no extremum at $x = 0$; local minimum at $x = 2$

5) Find and classify each of the critical points of the function $f(x) = \sin^2 x$ in the open interval $(-1, 3)$.

   Answer: global minimum at $x = 0$; global maximum at $x = \dfrac{\pi}{2}$

6) Find and classify each of the critical points of the function $f(x) = -\cos x - x\sin x$ in the open interval $(-3, 3)$.

   Answer: local maximum at $x = 0$; global minimums at $x = \pm\dfrac{\pi}{2}$

7) Determine two positive real numbers with product 2 in order to minimize the sum of their squares.

   Answer: $\sqrt{2}, \sqrt{2}$

8) What point on the parabola $y = x^2$ is closest to the point $(3, 0)$?

   Answer: $(1, 1)$

9) A rectangular box with square base and no top is to have a volume of exactly 1000 cm³. What is the minimum possible surface area of such a box?

   Answer: $300\sqrt[3]{4}$ cm²

10) Find the minimum possible value of the sum of a real number and its square.

   Answer: $-\dfrac{1}{4}$

11) What is the maximum possible volume of a right circular cylinder with total surface area 600π in.^{2} (including the top and the bottom)?

   Answer: 2000π in.^{3}

12) You need a cardboard container in the shape of a right circular cylinder of volume 54π in.^{2}. What radius r and height h would minimize its total surface area (including top and bottom)?

   Answer: r = 3 in., h = 6 in.

13) An aquarium has a square base made of slate costing 8¢/in.^{2} and four glass sides costing 3¢/in.^{2}. The volume of the aquarium is to be 36,000 in.^{3}. Find the dimensions of the least expensive such aquarium.

   Answer: 30 in. × 30 in. × 40 in.

14) The sum of two positive numbers is 160. What is the maximum possible value of their product?

   Answer: 6,400

15) A poster is to contain 96 in.^{2} of print, and each copy must have 3-in. margins at top and bottom and 2-in. margins on each side. What are the dimensions of such a poster having the least possible area?

   Answer: 12 in. × 18 in.

16) The pages of a book are each to contain 30 in.^{2} of print, and each page must have 2-in. margins at the top and bottom and 1-in. margins on each side. What are the dimensions of such a page having the least possible total area?

   Answer: $(\sqrt{15} + 2)$ in. × $(2\sqrt{15} + 4)$ in.

17) A storage building is to be shaped like a box with a square base (its floor). The floor costs $3/m^2, the walls cost $7/m^2, and the flat roof costs $5/m^2. The volume of the building is to be 12,544 m^3. What is the shape of the least expensive such building?

   Answer: 28 m × 28 m × 16 m

18) What is the maximum possible area of a rectangle inscribed in the ellipse $x^2 + 4y^2 = 4$ with the sides of the rectangle parallel to the coordinate axes?

   Answer: 4

19) A wall 8 m high stands 4 m away from a building. What is the length of the shortest ladder that will lean over the wall and touch the building? Use as independent variable the angle that the ladder makes with the ground.

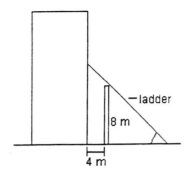

Answer: $4(1 + 2^{2/3})^{3/2}$ m

20) A rectangular sheet of thin metal is 5 m wide and 8 m long. Four small equal squares are cut from its corners and the projections of the resulting cross-shaped piece of metal are bent upward and welded to make an open-topped box with a rectangular base. What is the maximum possible volume of such a box?

Answer: 18 m³

21) You plan to build a playing field in the shape of a rectangle with a semicircular region at each end, so that races can be held around the perimeter of the field. If you want the total perimeter of the field to be 1000 m, what dimensions should your field have to maximize the area of the rectangular portion of the field?

Answer: The rectangular region should be $\frac{500}{\pi}$ m × 250 m, with the radius of each semicircular region being $\frac{250}{\pi}$ m.

4.5 Simple Curve Sketching
Solve the problem.

1) Use behavior "at infinity" to match the given functions with their graphs.

   1) $f(x) = x^3 - 4x - 1$ 2) $f(x) = x^4 - 2x^2 - x - \frac{1}{2}$ 3) $f(x) = -x^5 + 3x^2 - 2x + 2$ 4) $f(x) = -x^6 + x^5 + 2x^4 - 3x + \frac{3}{2}$

A)
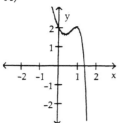

B)
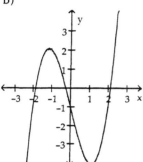

C)
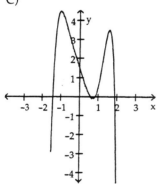

D)
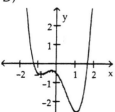

Answer: 1B, 2D, 3A, 4C

2) The graph of the function $y = f(x) = 2x^2 - 6x - 9$ is given below. Find any critical points and increasing/decreasing intervals for f(x).

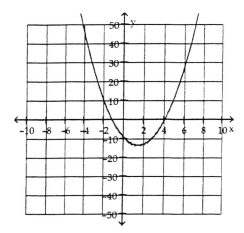

Answer: critical point at $x = \frac{3}{2}$; decreasing on $(-\infty, \frac{3}{2})$, increasing on $(\frac{3}{2}, \infty)$

3) The graph of the function $y = f(x) = 3x^3 + 3x^2 - 24x + 3$ is given below. Find any critical points and increasing/decreasing intervals for f(x).

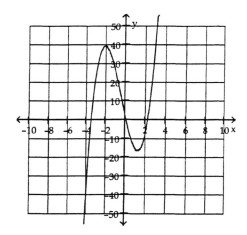

Answer: critical points at $x = -2$ and $\frac{4}{3}$; decreasing on $(-2, \frac{4}{3})$, increasing on $(-\infty, -2) \cup (\frac{4}{3}, \infty)$

4) The graph of the function $y = f(x) = -20 + 72x^2 + 32x^3 - 45x^4$ is given below. Find any critical points and increasing/decreasing intervals for f(x).

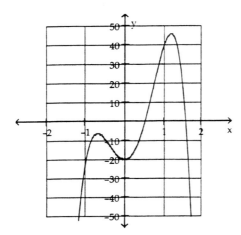

Answer: critical points at $x = -\frac{2}{3}, 0,$ and $\frac{6}{5}$; decreasing on $(-\frac{2}{3}, 0) \cup (\frac{6}{5}, \infty)$, increasing on $(-\infty, -\frac{2}{3}) \cup (0, \frac{6}{5})$

5) The graph of the function $y = f(x) = 3x^5 - 65x^3 + 540x + 5$ is given below. Find any critical points and increasing/decreasing intervals for f(x).

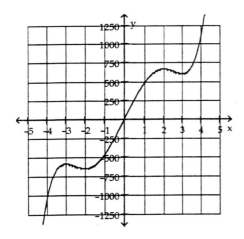

Answer: critical points at $x = \pm 2$ and ± 3; decreasing on $(-3, -2) \cup (2, 3)$, increasing on $(-\infty, -3) \cup (-2, 2) \cup (3, \infty)$

6) The graph of the function $y = f(x) = 3x^8 - 80x^6 + 384x^4 - 100$ is given below. Find any critical points and increasing/decreasing intervals for $f(x)$.

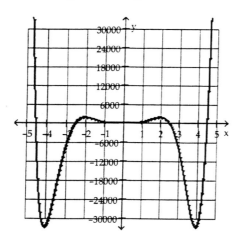

Answer: critical points at $x = \pm 2, \pm 4$, and 0; decreasing on $(-\infty, -4) \cup (-2, 0) \cup (2, 4)$, increasing on $(-4, -2) \cup (0, 2) \cup (4, \infty)$

7) Find the intervals on which the function $f(x) = 4x^5 - 5x^4$ is increasing and decreasing. Sketch the graph of $y = f(x)$, and identify any local maxima and minima. Any global extrema should also be identified.

Answer: local maximum at $(0, 0)$
local minimum at $(1, -1)$
increasing on $(-\infty, 0) \cup (1, \infty)$
decreasing on $(0, 1)$

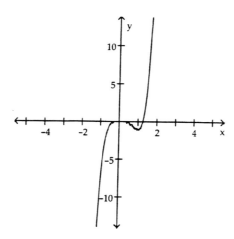

8) Find the interval on which the function $f(x) = 5x^2 + 10x - 7$ is increasing and decreasing. Sketch the graph of $y = f(x)$, and identify any local maxima and minima. Any global extrema should also be identified.

Answer: global minimum at $(-1, -12)$
increasing on $(-1, \infty)$
decreasing on $(-\infty, -1)$

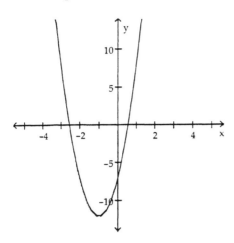

9) Find the interval on which the function $f(x) = (x - 2)^2(x + 3)^2$ is increasing and decreasing. Sketch the graph of $y = f(x)$, and identify any local maxima and minima. Any global extrema should also be identified.

Answer: global minima at $(-3, 0)$ and $(2, 0)$
local maximum at $(-\frac{1}{2}, \frac{625}{16})$
increasing on $(-3, -\frac{1}{2}) \cup (2, \infty)$
decreasing on $(-\infty, -3) \cup (-\frac{1}{2}, 2)$

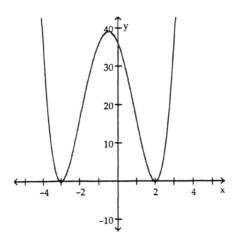

Calculus 83

10) Find the intervals on which the function $f(x) = 3\sqrt{x} - \frac{1}{4}x\sqrt{x}$ is increasing and decreasing. Sketch the graph of $y = f(x)$, and identify any local maxima and minima. Any global extrema should also be identified.

Answer: global maximum at (4, 4)
local minimum at (0, 0)
increasing on (0, 4)
decreasing on (4, ∞)

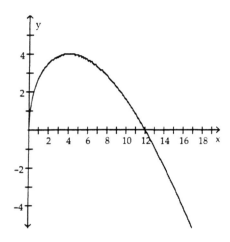

11) Find the intervals on which the function $f(x) = x^{2/3}(10 - x)$ is increasing and decreasing. Sketch the graph of $y = f(x)$, and identify any local maxima and minima. Any global extrema should also be identified.

Answer: local minimum at (0, 0) (and a cusp)
local maximum at $(4, 12\sqrt[3]{2})$
increasing on (0, 4)
decreasing on (-∞, 0) ∪ (4, ∞)

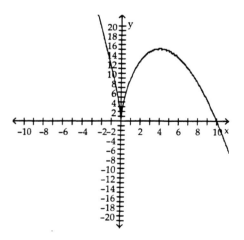

12) Find the intervals on which the function $f(x) = 3x^5 - 20x^3$ is increasing and decreasing. Sketch the graph of $y = f(x)$, and identify any local maxima and minima. Any global extrema should also be identified.

Answer: local minimum at $(2, -64)$
local maximum at $(-2, 64)$
increasing on $(-\infty, -2) \cup (2, \infty)$
decreasing on $(-2, 2)$ (but with a horizontal tangent at $(0, 0)$)

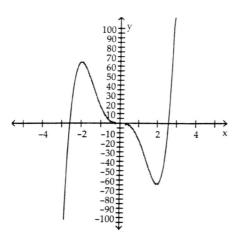

13) Find the intervals on which the function $f(x) = x(x - 1)^{4/5}$ is increasing and decreasing. Sketch the graph of $y = f(x)$, and identify any local maxima and minima. Any global extrema should also be identified.

Answer: local minimum (and a cusp) at $(1, 0)$
local maximum at $(\frac{5}{9}, \approx 0.2904)$
increasing on $(-\infty, \frac{5}{9}) \cup (1, \infty)$
decreasing on $(\frac{5}{9}, 1)$

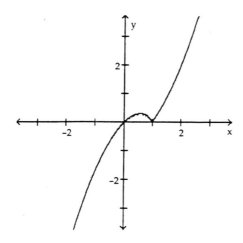

14) Find the intervals on which the function $f(x) = \dfrac{3}{x}$ is increasing and decreasing. Sketch the graph of of $y = f(x)$, and identify any local maxima and minima. Any global extrema should also be identified.

Answer: no extrema
decreasing on $(-\infty, 0) \cup (0, \infty)$

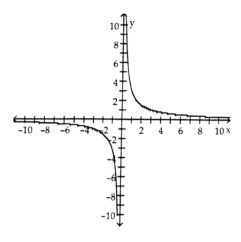

15) Find the interval on which the function $f(x) = x^4 + 4x^3$ is increasing and decreasing. Sketch the graph of $y = f(x)$, and identify any local maxima and minima. Any global extrema should also be identified.

Answer: global minimum at $(-3, -27)$
increasing on $(-3, 0) \cup (0, \infty)$
decreasing on $(-\infty, -3)$

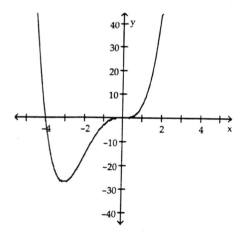

16) Find the intervals on which the function $f(x) = x^3 - 27x$ is increasing and decreasing. Sketch the graph of $y = f(x)$, and identify any local maxima and minima. Any global extrema should also be identified.

Answer: local maximum at $(-3, 54)$
local minimum at $(3, -54)$
increasing on $(-\infty, -3) \cup (3, \infty)$
decreasing on $(-3, 3)$

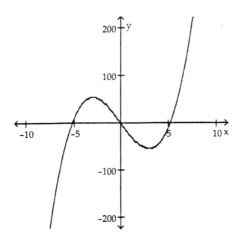

17) Find the intervals on which the function $f(x) = x^3 - 3x^2 - 9x$ is increasing and decreasing. Sketch the graph of $y = f(x)$, and identify any local maxima and minima. Any global extrema should also be identified.

Answer: local maximum at $(-1, 5)$
local minimum at $(3, -27)$
increasing on $(-\infty, -1) \cup (3, \infty)$
decreasing on $(-1, 3)$

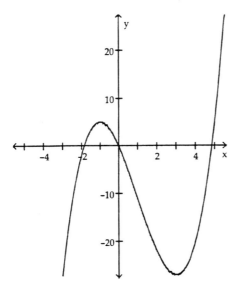

18) Find the intervals on which the function f(x) = $-x^3 + 6x^2 - 10$ is increasing and decreasing. Sketch the graph of y = f(x), and identify any local maxima and minima. Any global extrema should also be identified.

Answer: local minimum at (0, -10)
local maximum at (4, 22)
increasing on (0, 4)
decreasing on $(-\infty, 0) \cup (4, \infty)$

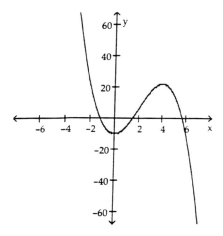

19) Find the intervals on which the function f(x) = $x^4 - 2x^2 + 1$ is increasing and decreasing. Sketch the graph of y = f(x), and identify any local maxima and minima. Any global extrema should also be identified.

Answer: local maximum at (0, 1)
global minima at (-1, 0) and (1, 0)
increasing on $(-1, 0) \cup (1, \infty)$
decreasing on $(-\infty, -1) \cup (0, 1)$

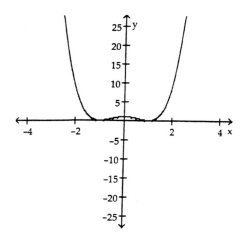

20) The values of the function f(x) at its critical points are given, together with the graph y = f'(x). Use this information to construct a graph y = f(x) of the function.

f(−3) = 86, f(1) = −10

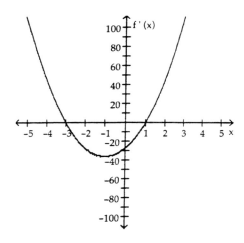

Answer: f(x) = 3x³ + 9x² − 27x + 5

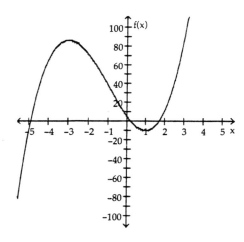

21) The values of the function f(x) at its critical points are given, together with the graph y = f'(x). Use this information to construct a graph y = f(x) of the function.

$f(-3) = -4$, $f(-1) = 28$, $f(4) = -347$

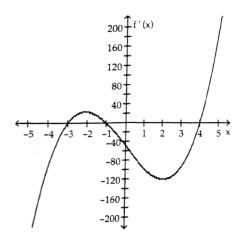

Answer: $f(x) = x^4 - 26x^2 - 48x + 5$

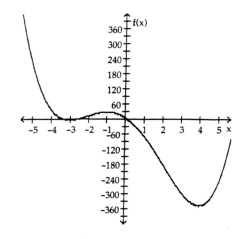

4.6 Higher Derivatives and Concavity
Solve the problem.

1) Find the first three derivatives of the function $f(x) = 3x^4 - 4x^3 + 5x + 1$.

   Answer: $f'(x) = 12x^3 - 12x^2 + 5$
   $f''(x) = 36x^2 - 24x$
   $f'''(x) = 72x - 24$

2) Find the first three derivatives of the function $g(t) = (4t - 5)^{3/2}$.

   Answer: $f'(x) = 6\sqrt{4t - 5}$
   $f''(x) = \dfrac{12}{\sqrt{4t - 5}}$
   $f'''(x) = \dfrac{-24}{(4t - 5)^{3/2}}$

Calculus 90

3) Find the first three derivatives of the function f(x) = 2cos x sin 2x.

   Answer: f'(x) = 3cos 3x + cos x
   f''(x) = −9sin 3x − sin x
   f'''(x) = −27cos 3x − cos x

4) Calculate $\dfrac{dy}{dx}$ and $\dfrac{d^2y}{dx^2}$ of the implicitly defined function: $x^2 + y^2 = 10$.

   Answer: $\dfrac{dy}{dx} = \dfrac{-x}{y}$

   $\dfrac{d^2y}{dx^2} = \dfrac{-(x^2 + y^2)}{y^3}$

5) Calculate $\dfrac{dy}{dx}$ and $\dfrac{d^2y}{dx^2}$ of the implicitly defined function: $3x^2 + 5xy + 3y^2 = 6$.

   Answer: $\dfrac{dy}{dx} = \dfrac{-(6x + 5y)}{5x + 6y}$

   $\dfrac{d^2y}{dx^2} = \dfrac{-22(3x^2 + 5xy + 3y^2)}{(5x + 6y)^3}$

6) Calculate $\dfrac{dy}{dx}$ and $\dfrac{d^2y}{dx^2}$ of the implicitly defined function: cos y = xy.

   Answer: $\dfrac{dy}{dx} = \dfrac{y}{-\sin y - x}$

   $\dfrac{d^2y}{dx^2} = \dfrac{y(2x - y\cos y + 2\sin y)}{(x + \sin y)^3}$

7) Calculate $\dfrac{dy}{dx}$ and $\dfrac{d^2y}{dx^2}$ of the implicitly defined function: $\cos^2 x + \sin^2 y = 1$.

   Answer: $\dfrac{dy}{dx} = \dfrac{\sin 2x}{\sin 2y}$

   $\dfrac{d^2y}{dx^2} = \dfrac{2\sin 2y \cos 2x - 2\sin^2 2x \cot 2y}{\sin^2 2y}$

8) Find the exact coordinates of the inflection points and critical points marked on the graph of $f(x) = \frac{2}{3}x^3 - 2x^2 - 6x$.

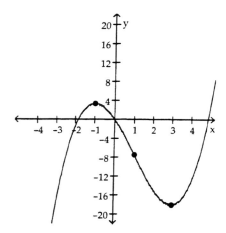

Answer: maximum at $(-1, \frac{10}{3})$; minimum at $(3, -18)$; inflection point at $(1, -\frac{22}{3})$

9) Find the exact coordinates of the inflection points and critical points marked on the graph of $f(x) = 2x^3 + 3x^2 - 180x + 150$.

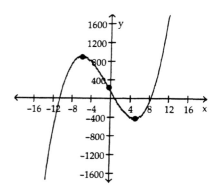

Answer: maximum at $(-6, 906)$; minimum at $(5, -425)$; inflection point at $(-\frac{1}{2}, \frac{481}{2})$

Calculus 92

10) Find the exact coordinates of the inflection points and critical points marked on the graph of
$f(x) = 4x^5 - 100x^3 - 5$.

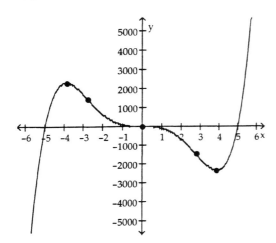

Answer: maximum at $(-\sqrt{15}, \approx 2318.79)$; minimum at $(\sqrt{15}, \approx -2328.79)$; inflection points at $(0, -5)$, $(\frac{\sqrt{30}}{2}, \approx -1442.77)$, $(-\frac{\sqrt{30}}{2}, \approx 1432.77)$

11) Apply the second derivative test to find the local and global extrema and the points of inflection of
$f(x) = 7 - 6x - x^2$.

Answer: global and local maximum at $(-3, 16)$; no points of inflection

12) Apply the second derivative test to find the local and global extrema and the points of inflection of
$f(x) = \cos 2x$, $-\pi \leq x \leq \pi$.

Answer: global and local maximums at $(0, 1)$ and $(\pm \pi, 1)$; global and local minimums at $(\pm \frac{\pi}{2}, -1)$; points of inflection at $(\pm \frac{\pi}{4}, 0)$ and $(\pm \frac{3\pi}{4}, 0)$

13) Sketch the graph of the function, indicating all critical points and inflection points. Apply the second derivative test at each critical point. Show the correct concave structure and indicate the behavior of f(x) as $x \to \pm \infty$.

$f(x) = 4x^5 - 5x^4$

Answer: maximum at $(0, 0)$
minimum at $(1, -1)$
inflection point at $(\frac{3}{4}, \frac{-81}{128})$
increasing on $(-\infty, 0) \cup (1, \infty)$
decreasing on $(0, 1)$
concave up on $(\frac{3}{4}, \infty)$
concave down on $(-\infty, \frac{3}{4})$

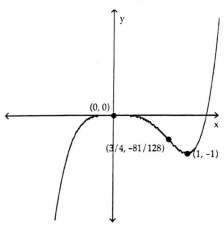

14) Sketch the graph of the function, indicating all critical points and inflection points. Apply the second derivative test at each critical point. Show the correct concave structure and indicate the behavior of f(x) as $x \to \pm \infty$.

$f(x) = x^4 + 4x^3$

Answer: minimum at (-3, -27)
inflection points at (-2, -16) and (0, 0)
increasing on (-3, ∞)
decreasing on (-∞, -3)
concave up on (-∞, -2) $\cup$ (0, ∞)
concave down on (-2, 0)

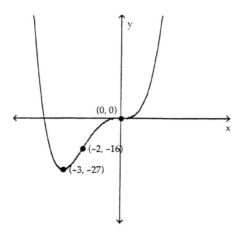

15) Sketch the graph of the function, indicating all critical points and inflection points. Apply the second derivative test at each critical point. Show the correct concave structure and indicate the behavior of f(x) as $x \to \pm \infty$.

$f(x) = x^3 - 27x$

Answer: maximum at (-3, 54)
minimum at (3, -54)
inflection point at (0, 0)
increasing on (-∞, -3) $\cup$ (3, ∞)
decreasing on (-3, 3)
concave up on (0, ∞)
concave down on (-∞, 0)

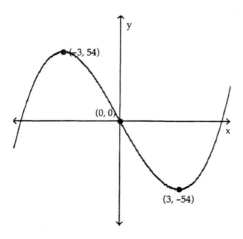

16) Sketch the graph of the function, indicating all critical points and inflection points. Apply the second derivative test at each critical point. Show the correct concave structure and indicate the behavior of f(x) as $x \to \pm \infty$.

$f(x) = \dfrac{x^2}{x-1}$

Answer: maximum (0, 0)
minimum (2, 4)
inflection point: NONE
increasing on $(-\infty, 0) \cup (2, \infty)$
decreasing on $(0, 1) \cup (1, 2)$
concave up on $(1, \infty)$
concave down on $(-\infty, 1)$
discontinuous at $x = 1$

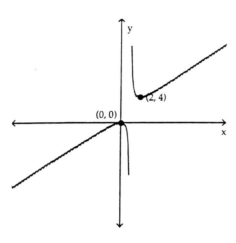

17) Sketch the graph of the function, indicating all critical points and inflection points. Apply the second derivative test at each critical point. Show the correct concave structure and indicate the behavior of $f(x)$ as $x \to \pm \infty$.

$f(x) = \dfrac{x^2}{x^2 - 1}$

Answer: maximum $(0, 0)$
inflection point: NONE
increasing on $(-\infty, -1) \cup (-1, 0)$
decreasing on $(0, 1) \cup (1, \infty)$
concave up on $(-\infty, -1) \cup (1, \infty)$
concave down on $(-1, 1)$
discontinuous at $x = \pm 1$

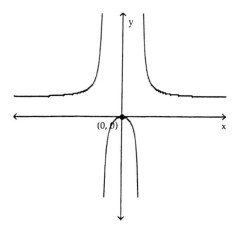

18) Sketch the graph of the function, indicating all critical points and inflection points. Apply the second derivative test at each critical point. Show the correct concave structure and indicate the behavior of $f(x)$ as $x \to \pm \infty$.

$f(x) = \sqrt[3]{x}$

Answer: maximum: NONE
minimum: NONE
inflection point at $(0, 0)$
increasing on $(-\infty, \infty)$
concave up on $(-\infty, 0)$
concave down on $(0, \infty)$

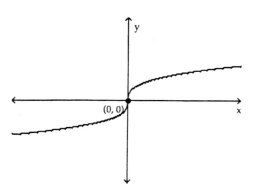

19) Sketch the graph of the function, indicating all critical points and inflection points. Apply the second derivative test at each critical point. Show the correct concave structure and indicate the behavior of f(x) as $x \to \pm \infty$.

$f(x) = x^{2/3}$

Answer: minimum at (0, 0) (cusp)
maximum: NONE
point of inflection: NONE
increasing on $(0, \infty)$
decreasing on $(-\infty, 0)$
concave down on $(-\infty, \infty)$

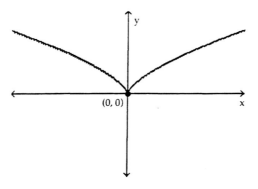

20) Sketch the graph of the function, indicating all critical points and inflection points. Apply the second derivative test at each critical point. Show the correct concave structure and indicate the behavior of f(x) as $x \to \pm \infty$.

$f(x) = x^3 - 3x^2 - 9x$

Answer: maximum at (-1, 5)
minimum at (3, -27)
inflection point at (1, -11)
increasing on $(-\infty, -1) \cup (3, \infty)$
decreasing on (1, 3)
concave down on $(-\infty, 1)$
concave up on $(1, \infty)$

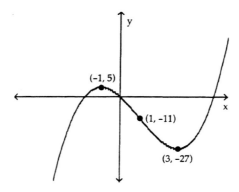

21) Sketch the graph of the function, indicating all critical points and inflection points. Apply the second derivative test at each critical point. Show the correct concave structure and indicate the behavior of f(x) as $x \to \pm \infty$.

$f(x) = x^4 - 8x^2 + 7$

Answer: maximum at (0, 7)

minimum at (± 2, -9)

inflection points at $(\pm \frac{2\sqrt{3}}{3}, -\frac{17}{9})$ {shown as a and b on graph}

increasing on (-2, 0) ∪ (2, ∞)
decreasing on (-∞, 2) ∪ (0, 2)
concave down on $(-\frac{2\sqrt{3}}{3}, \frac{2\sqrt{3}}{3})$
concave up on $(-\infty, -\frac{2\sqrt{3}}{3}) \cup (\frac{2\sqrt{3}}{3}, \infty)$

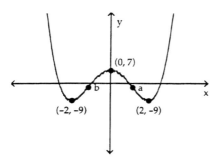

22) Sketch the graph of the function, indicating all critical points and inflection points. Apply the second derivative test at each critical point. Show the correct concave structure and indicate the behavior of f(x) as $x \to \pm \infty$.

$f(x) = \frac{x^2 - 1}{x^2 + 1}$

Answer: Maximum: NONE

minimum at (0, -1)

inflection points at $(\pm \frac{\sqrt{3}}{3}, -\frac{1}{2})$ {shown as a and b on graph}

increasing on (0, ∞)
decreasing on (-∞, 0)
concave down on $(-\infty, -\frac{\sqrt{3}}{3}) \cup (\frac{\sqrt{3}}{3}, \infty)$
concave up on $(-\frac{\sqrt{3}}{3}, \frac{\sqrt{3}}{3})$

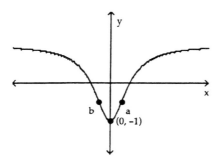

23) Sketch the graph of the function, indicating all critical points and inflection points. Apply the second derivative test at each critical point. Show the correct concave structure and indicate the behavior of f(x) as $x \to \pm\infty$.

$f(x) = x + \dfrac{1}{x}$

Answer: maximum at (−1, −2)
minimum at (1, 2)
inflection point: NONE
increasing on (−∞, 1) ∪ (1, ∞)
decreasing on (−1, 0) ∪ (0, 1)
concave down on (−∞, 0)
concave up on (0, ∞)
discontinuous at x = 0

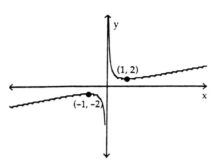

24) Match the graph of the function (1 - 5) with the graph of its second derivative (A-E).

(1)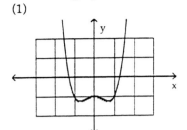

(A)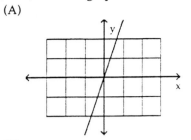

(2)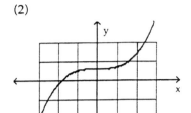

(B)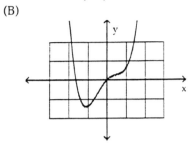

(3)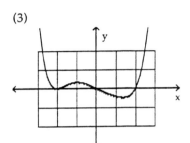

(C)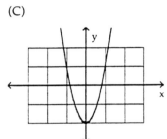

(4)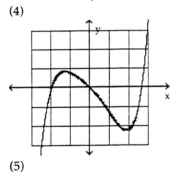

(D)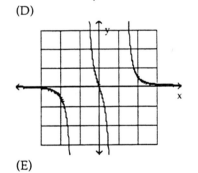

(5)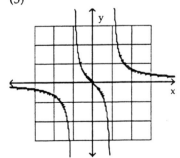

(E)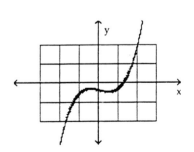

Answer: 1 C, 2 A, 3 B, 4 E, 5 D

4.7 Curve Sketching and Asymptotes
Solve the problem.

1) Find $\lim\limits_{x \to \infty} \dfrac{6x^4 - 5x^3 - 2x + 1}{3x^4 - 2x - 7}$.

   Answer: 2

2) Find $\lim\limits_{x \to -\infty} (5x - \sqrt{6x^2 - x})$.

   Answer: $\dfrac{19\sqrt{6}}{6}$

3) Find $\lim\limits_{x \to \infty} \dfrac{3x^2}{8x^2 + 4x}$.

   Answer: $\dfrac{3}{8}$

4) Apply your knowledge of limits and asymptotes to match each function with its graph. (Do not graph on a calculator)

   (1) $f(x) = \dfrac{4}{x^2 - 4}$ (2) $f(x) = \dfrac{1}{x + 1}$ (3) $f(x) = \dfrac{x^2}{1 - x}$ (4) $f(x) = \dfrac{x^3}{1 - x^2}$

   (A) 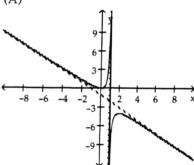   (B)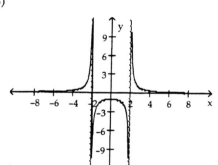

   (C) 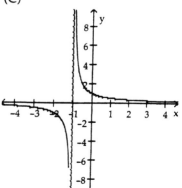   (D)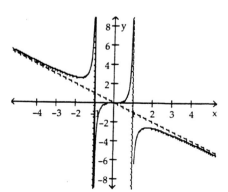

   Answer: 1 B, 2 C, 3 A, 4 D

5) Sketch, by hand, the graph of f(x). Identify all extrema, inflection points, intercepts, and asymptotes. Show the concave structure clearly and note any discontinuities.

$$f(x) = \frac{x^2}{x-1}$$

Answer: maximum at (0, 0)
minimum at (2, 4)
inflection point: NONE
x-intercept(s): (0, 0)
y-intercept(s): (0, 0)
increasing on $(-\infty, 0) \cup (2, \infty)$
decreasing on $(0, 1) \cup (1, 2)$
concave up on $(1, \infty)$
concave down on $(-\infty, 1)$
discontinuity at x = 1
asymptotes at x = 1 and y = x + 1

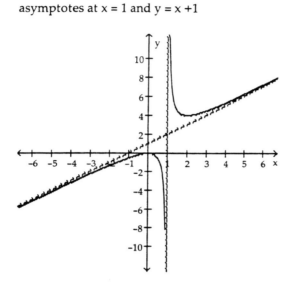

6) Sketch, by hand, the graph of f(x). Identify all extrema, inflection points, intercepts, and asymptotes. Show the concave structure clearly and note any discontinuities.

$$f(x) = \frac{x^2}{x^2 - 1}$$

Answer: maximum at (0, 0)
minimum: NONE
inflection point: NONE
x-intercept(s): (0, 0)
y-intercept(s): (0, 0)
increasing on $(-\infty, -1) \cup (-1, 0)$
decreasing on $(0, 1) \cup (1, \infty)$
concave up on $(-\infty, -1) \cup (1, \infty)$
concave down on $(-1, 1)$
discontinuity at $x = \pm 1$
asymptotes at $x = \pm 1$ and $y = 1$

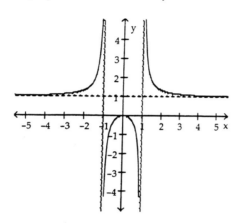

7) Sketch, by hand, the graph of f(x). Identify all extrema, inflection points, intercepts, and asymptotes. Show the concave structure clearly and note any discontinuities.

$$f(x) = \frac{x^2 - 1}{x^2 + 1}$$

Answer: minimum at (0, -1)
maximum: NONE
inflection points at (± 1, 0)
x-intercept(s): (± 1, 0)
y-intercept(s): (0, -1)
increasing on (0, ∞)
decreasing on (-∞, 0)
concave up on (-1, 1)
concave down on (-∞, -1) ∪ (1, ∞)
asymptote at y = 1

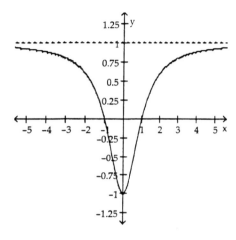

8) Sketch, by hand, the graph of f(x). Identify all extrema, inflection points, intercepts, and asymptotes. Show the concave structure clearly and note any discontinuities.

$f(x) = \dfrac{x}{x^2 + 1}$

Answer: maximum at $(1, \frac{1}{2})$

minimum at $(-1, -\frac{1}{2})$

inflection points at $(0, 0)$, $(\sqrt{3}, \frac{\sqrt{3}}{4})$, and $(-\sqrt{3}, -\frac{\sqrt{3}}{4})$

x-intercept(s): (0, 0)
y-intercept(s): (0, 0)
increasing on (−1, 1)
decreasing on (−∞, −1) ∪ (1, ∞)
concave up on $(-\sqrt{3}, 0) \cup (\sqrt{3}, \infty)$
concave down on $(-\infty, -\sqrt{3}) \cup (0, \sqrt{3})$
asymptote at y = 0

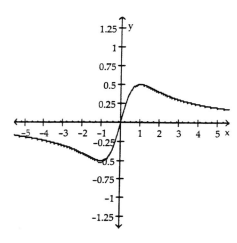

9) Sketch, by hand, the graph of f(x). Identify all extrema, inflection points, intercepts, and asymptotes. Show the concave structure clearly and note any discontinuities.

$f(x) = x + \dfrac{1}{x}$

Answer: maximum at $(-1, -2)$
minimum at $(1, 2)$
inflection point: NONE
x-intercept(s): NONE
y-intercept(s): NONE
increasing on $(-\infty, -1) \cup (1, \infty)$
decreasing on $(-1, 0) \cup (0, 1)$
concave up on $(0, \infty)$
concave down on $(-\infty, 0)$
asymptotes at $y = x$ and $x = 0$
discontinuous at $x = 0$

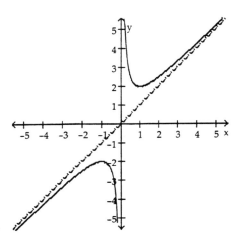

10) Sketch, by hand, the graph of f(x). Identify all extrema, inflection points, intercepts, and asymptotes. Show the concave structure clearly and note any discontinuities.

$$f(x) = \frac{x}{x^2 - 1}$$

Answer: maximum: NONE
minimum: NONE
inflection point at (0, 0)
x-intercept(s): (0, 0)
y-intercept(s): (0, 0)
decreasing on $(-\infty, -1) \cup (-1, 1) \cup (1, \infty)$
concave up on $(-1, 0) \cup (1, \infty)$
concave down on $(-\infty, -1) \cup (0, 1)$
asymptotes at $y = 0$ and $x = \pm 1$
discontinuous at $x = \pm 1$

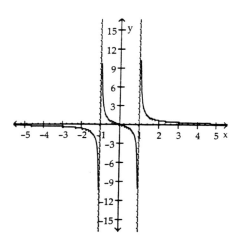

11) Sketch, by hand, the graph of f(x). Identify all extrema, inflection points, intercepts, and asymptotes. Show the concave structure clearly and note any discontinuities.

$$f(x) = \frac{x+3}{x-3}$$

Answer: maximum: NONE
minimum: NONE
inflection point: NONE
x-intercept(s): (-3, 0)
y-intercept(s): (0, -1)
decreasing on $(-\infty, 3) \cup (3, \infty)$
concave up on $(3, \infty)$
concave down on $(-\infty, 3)$
asymptotes: x = 3 and y = 1
discontinuous at x = 3

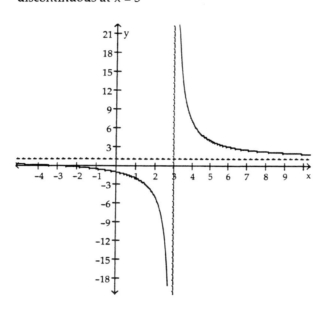

12) Sketch, by hand, the graph of f(x). Identify all extrema, inflection points, intercepts, and asymptotes. Show the concave structure clearly and note any discontinuities.

$$f(x) = \frac{x}{4 - x^2}$$

Answer: maximum: NONE
minimum: NONE
inflection point at (0, 0)
x-intercept(s): (0, 0)
y-intercept(s): (0, 0)
increasing on $(-\infty, -2) \cup (-2, 2) \cup (2, \infty)$
concave up on $(-\infty, -2) \cup (0, 2)$
concave down on $(-2, 0) \cup (2, \infty)$
asymptotes: $x = \pm 2$ and $y = 0$
discontinuous at $x = \pm 2$

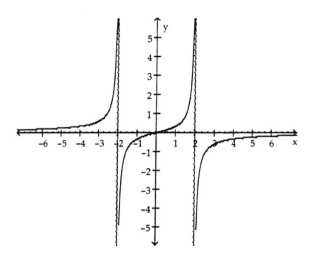

13) Sketch, by hand, the graph of f(x). Identify all extrema, inflection points, intercepts, and asymptotes. Show the concave structure clearly and note any discontinuities.

$$f(x) = \frac{4(x-1)}{x^2+1}$$

Answer: maximum at $(1+\sqrt{2}, 2\sqrt{2} - 2) \approx (2.41421, 0.828427)$
minimum at $(1-\sqrt{2}, -2\sqrt{2} - 2) \approx (-0.414214, -4.82843)$
inflection points at $(-1, -4)$, $(2-\sqrt{3}, -\sqrt{3} - 1)$ and $(2+\sqrt{3}, \sqrt{3} - 1)$
x-intercept(s): $(1, 0)$
y-intercept(s): $(0, -4)$
increasing on $(1-\sqrt{2}, 1+\sqrt{2})$
decreasing on $(-\infty, 1-\sqrt{2}) \cup (1+\sqrt{2}, \infty)$
concave up on $(-1, 2-\sqrt{3}) \cup (2+\sqrt{3}, \infty)$
concave down on $(-\infty, -1) \cup (2-\sqrt{3}, 2+\sqrt{3})$
asymptote: $y = 0$

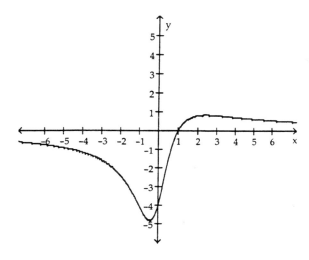

Ch. 5 The Integral

5.1 Introduction

1) There are no exercises for this section.

   Answer:

5.2 Antiderivatives and Initial Value Problems

Solve the problem.

1) $\int x^5 \sin x^6 \, dx$

   Answer: $-\frac{1}{6}\cos x^6 + C$

2) $\int \sin^4 x \cos x \, dx$

   Answer: $\frac{1}{5}\sin^5 x + C$

3) $\int 2x \cos x^2 \, dx$

   Answer: $\sin x^2 + C$

4) $\int x^2 (x^3 + 1)^{10} \, dx$

   Answer: $\frac{1}{33}(x^3 + 1)^{11} + C$

5) $\int \frac{x+1}{x^5} \, dx$

   Answer: $-\frac{1}{4x^4} - \frac{1}{3x^3} + C$

6) $\int \sqrt{x} \cos x\sqrt{x} \, dx$

   Answer: $\frac{2}{3}\sin x\sqrt{x} + C$

7) $\int \sin x \cos^7 x \, dx$

   Answer: $-\frac{1}{8}\cos^8 x + C$

8) $\int \left(x - \frac{1}{x}\right)^2 dx$

   Answer: $\frac{1}{3}x^3 - 2x - \frac{1}{x} + C$

9) $\int (\cos 2x - \sin 2x) \, dx$

   Answer: $\frac{1}{2}(\cos 2x + \sin 2x) + C$

10) $\int x^5 (x^6 - 1)^7 \, dx$

    Answer: $\frac{1}{48}(x^6 - 1)^8 + C$

11) $\int t \cos t^2 \, dt$

Answer: $\dfrac{1}{2}\sin t^2 + C$

12) $\int (7u^3 + 7) \, du$

Answer: $\dfrac{7}{4} u^4 + 7u + C$

13) $\int (x+1)^5 \, dx$

Answer: $\dfrac{1}{6}(x+1)^6 + C$

14) $\int \cos 7x \, dx$

Answer: $\dfrac{1}{7} \sin 7x + C$

15) $\int \sin t \cos t \, dt$

Answer: $-\dfrac{1}{2}\cos^2 t + C$ OR $\dfrac{1}{2}\sin^2 t + C$

16) $\int 3x^2 \cos x^3 \, dx$

Answer: $\sin x^3 + C$

17) $\int \dfrac{\cos x}{\sin^3 x} \, dx$

Answer: $-\dfrac{1}{2} \csc^2 x + C$

18) $\int \dfrac{x+1}{\sqrt{x}} \, dx$

Answer: $2\sqrt{x} + \dfrac{2}{3} x^{3/2} + C$

19) $\int \dfrac{2x}{\sqrt{x^2+1}} \, dx$

Answer: $2\sqrt{x^2+1} + C$

20) $\int \dfrac{7}{(x+1)^2} \, dx$

Answer: $-\dfrac{7}{x+1} + C$

21) $\int \dfrac{x^3+4}{x^2} \, dx$

Answer: $\dfrac{1}{2}x^2 - \dfrac{4}{x} + C$

22) $\int \dfrac{2x^2}{\sqrt{x^3+5}}\, dx$

Answer: $\dfrac{4}{3}\sqrt{x^3+5} + C$

23) $\int x^2(x^7 - x^5)\, dx$

Answer: $\dfrac{1}{10}x^{10} - \dfrac{1}{8}x^8 + C$

24) $\int (\sin x)^{99} \cos x\, dx$

Answer: $\dfrac{1}{100}\sin^{100} x + C$

25) $\int \sqrt{x+1}\, dx$

Answer: $\dfrac{2}{3}(x+1)^{3/2} + C$

26) $\int z^2(z^3+1)^{1/3}\, dz$

Answer: $\dfrac{(z^3+1)^{4/3}}{4} + C$

27) $\int x^4(x^5+7)^{99}\, dx$

Answer: $\dfrac{1}{500}(x^5+7)^{100} + C$

28) $\int \sin 2x \sin x\, dx$

Answer: $-\dfrac{1}{2}\sin x - \dfrac{1}{6}\sin 3x + C$

29) $\int x\sqrt{1-x^2}\, dx$

Answer: $-\dfrac{(1-x^2)^{3/2}}{3} + C$

30) $\int 2t \sin t^2\, dt$

Answer: $-\cos t^2 + C$

31) $\int \dfrac{1}{x^2} \cos \dfrac{1}{x}\, dx$

Answer: $-\sin \dfrac{1}{x} + C$

32) $\int \dfrac{1}{\sqrt{x}} \cos \sqrt{x}\, dx$

Answer: $2\sin \sqrt{x} + C$

33) Solve the initial value problem: $\frac{dy}{dx} = 2x + 1$; $y(0) = 3$

    Answer: $y = x^2 + x + 3$

34) Solve the initial value problem: $\frac{dy}{dx} = (x - 2)^3$; $y(2) = 1$

    Answer: $y = \frac{1}{4}(x - 2)^4 + 1$

35) Solve the initial value problem: $\frac{dy}{dx} = \sqrt{x}$; $y(4) = 0$

    Answer: $y = \frac{2}{3}x^{3/2} - \frac{16}{3}$

36) Solve the initial value problem: $\frac{dy}{dx} = \frac{1}{x^2}$; $y(1) = 5$

    Answer: $y = 6 - \frac{1}{x}$

37) Solve the initial value problem: $\frac{dy}{dx} = \frac{1}{\sqrt{x+2}}$; $y(2) = -1$

    Answer: $y = 2\sqrt{x+2} - 5$

38) Solve the initial value problem: $\frac{dy}{dx} = x\sqrt{9 + x^2}$; $y(-4) = 0$

    Answer: $y = \frac{1}{3}(x^2 + 9)^{3/2} - \frac{125}{3}$

39) Solve the initial value problem: $\frac{dy}{dx} = \cos 2x$; $y(0) = 1$

    Answer: $y = \frac{1}{2}\sin 2x + 1$

40) Solve the initial value problem: $\frac{dy}{dx} = \sec x \tan x$; $y(0) = 2$

    Answer: $y = \sec x + 1$

41) A particle moves along the x-axis with the acceleration function $a(t)$, initial position $x(0)$, and initial velocity $v(0)$. Find the particle's position function $x(t)$.
    $a(t) = 10t - 6$ $x(0) = 5$; $v(0) = 0$

    Answer: $x(t) = \frac{5t^3}{3} - 3t^2 + 5$

42) A particle moves along the x-axis with the acceleration function $a(t)$, initial position $x(0)$, and initial velocity $v(0)$. Find the particle's position function $x(t)$.
    $a(t) = 25 - 10t$ $x(0) = 5$; $v(0) = 10$

    Answer: $x(t) = \frac{25t^2}{2} - \frac{5t^3}{3} + 10t + 5$

43) A particle moves along the x-axis with the acceleration function $a(t)$, initial position $x(0)$, and initial velocity $v(0)$. Find the particle's position function $x(t)$.

$a(t) = t^2 - 4t + 6$ $\quad x(0) = 4;\ v(0) = 1$

Answer: $x(t) = \dfrac{t^4}{12} - \dfrac{2t^3}{3} + 3t^2 + t + 4$

44) A particle moves along the x-axis with the acceleration function $a(t)$, initial position $x(0)$, and initial velocity $v(0)$. Find the particle's position function $x(t)$.

$a(t) = 3t^2$ $\quad x(0) = -3;\ v(0) = 12$

Answer: $x(t) = \dfrac{t^4}{4} + 12t - 3$

45) A particle moves along the x-axis with the acceleration function $a(t)$, initial position $x(0)$, and initial velocity $v(0)$. Find the particle's position function $x(t)$.

$a(t) = 12 + 3\sqrt{t}$ $\quad x(0) = 4;\ v(0) = -3$

Answer: $x(t) = 6t^2 + \dfrac{4t^{5/2}}{5} - 3t + 4$

46) A particle moves along the x-axis with the acceleration function $a(t)$, initial position $x(0)$, and initial velocity $v(0)$. Find the particle's position function $x(t)$.

$a(t) = 5\sin 5t$ $\quad x(0) = -1;\ v(0) = 2$

Answer: $x(t) = \dfrac{-\sin 5t}{5} + 3t - 1$

47) An ostrich egg is dropped from an altitude of 14,400 ft above the earth's surface. Ignore air resistance, use $g = 32$ ft/s^2 to find both how long the egg stays aloft, and its impact velocity.

Answer: stays in the air 30 seconds
$v(t) = 960$ ft/s

48) A rock is dropped from a balloon hovering at 4800 ft above the ground. Neglect air resistance, use $g = 32$ ft/s^2, to find both how long it takes for the rock to reach the ground and its impact velocity.

Answer: stays in the air $10\sqrt{3}$ seconds
$v(10\sqrt{3}) = 320\sqrt{3}$ ft/s

49) A punter kicked a football straight upward and the sportscaster reported that the ball had a "hang time" of 4.0 seconds. How high did the ball go, and what was its initial velocity?

Answer: $h = 64$ ft
$v = 64$ ft/s

50) An arrow shot vertically upward, reached a maximum height of 1920 ft. What was its initial velocity?

Answer: $v = 64\sqrt{30}$ ft/s

51) A ball is thrown upward with initial velocity of 64 ft/s. How high will it go?

Answer: $h = 64$ ft

52) A stone dropped from a height of 1 ft near the north pole of the planet Mesklin strikes the ground 0.001 seconds later. Find the value of the acceleration g of gravity on the planet, and also find the impact velocity of the stone.

Answer: $g = 2,000,000$ ft/s
$v = 2000$ ft/s

53) A ball is thrown upward with initial velocity of 144 ft/s. How high will it go?

Answer: $h = 324$ ft

54) A stone is dropped from the top of a building 960 ft high. How long will it remain aloft? What will its impact velocity be?

Answer: $t = 2\sqrt{15}$ s
$v = 64\sqrt{15}$ ft/s

55) A crossbow bolt was shot upward remained aloft for 48 s. What was its initial velocity, and how high did it go?

Answer: $v = 768$ ft/s (An unusually powerful crossbow!)
$h = 9216$ ft

56) A stone dropped into a well hits the bottom 6 s later. How deep is the well?

Answer: $h = 576$ ft

57) A croquet ball is thrown upward from the top of a building 960 ft high with an initial velocity of 48 ft/s. How high will it go? How long will it remain aloft?

Answer: $h = 996$ ft
t (back to top of building) = 3 s
t (to ground) = $(3 + \sqrt{249})/2$ s

58) A watermelon is thrown vertically upward from the ground. Neglect air resistance and take the acceleration of gravity to be 32 ft/s². With what initial velocity must the watermelon be thrown so that its maximum altitude is 4096 ft? In such a case, how long will the watermelon remain aloft?

Answer: $v = 512$ ft/s
$t = 32$ s

59) A very dense planet has $g = 15,000$ ft/s² for the acceleration of gravity at its surface. If a stone is dropped there from an altitude of 3 ft, how long will it remain aloft? With what speed will the stone strike the ground?

Answer: $t = 1/50$ s
$v = 300$ ft/s

60) When its brakes are applied, a certain automobile decelerates at 11 ft/s². Suppose that its initial velocity is 60 mi/h (88 ft/s). How long will the automobile travel before coming to a stop? How far will it travel?

Answer: $t = 8$ s
$d = 352$ ft

61) An arrow is shot upward from the surface of the moon, where acceleration of gravity is 5.2 ft/s². The initial velocity of the arrow is 1040 ft/s. How high will the arrow go, and how long will it remain aloft?

Answer: $h = 104,000$ ft
$t = 400$ s

62) A zucchini is dropped from the top of a tall building. Its speed at impact is 256 ft/s. Ignore air resistance, use $g = 32$ ft/s^2, to find the height of the building.

Answer: $h = 1024$ ft

63) When the space shuttle first leaves the ground, its acceleration is a (constant) 60 ft/s^2. How long does it take the shuttle to reach a height of 3 mi (under the assumption that it climbs vertically)? How fast is it traveling then?

Answer: $t = 4\sqrt{33}$ s
$v = 240\sqrt{33}$ ft/s

64) In a panic stop, a certain car leaves skid marks 420 ft long. This car is known to have constant deceleration under such circumstances; in an experiment under similar conditions the car left skid marks 105 ft long when the brakes were applied at 30 ft/s. How fast was the car moving when the brakes were applied in the panic stop?

Answer: $v = 60$ ft/s

65) A stone is thrown vertically upward from the surface of a very small planet where the value of g, the acceleration of gravity, is 5 ft/s^2. The initial velocity of the stone is 100 ft/s. How high does the stone go, and how long does it remain aloft?

Answer: $h = 1000$ ft
$t = 40$ s

66) A ball thrown vertically upward from the earth's surface reaches a maximum altitude of 225 ft. Use $g = 32$ ft/s^2, ignore air resistance, and compute the initial velocity of the ball.

Answer: $v = 120$ ft/s

67) The maximum deceleration of a certain freight train is 1 ft/s^2. If it is traveling at 60 mi/h, what is the minimum time in which it can come to a stop? How far will it travel during that time?

Answer: $t = 88$ s
$d = 3872$ ft

68) On March 6, 1992, Schroeder told Charlie Brown that a baseball dropped from the height of Charlie's waist would strike the ground at a speed of 9.45 mi/h. How high, to the nearest inch, is Charlie Brown's waist?

Answer: $h = 3$ ft, 0 inches

69) A particle is traveling with acceleration $a(t) = \sqrt{4 + 3t}$ ft/s^2. At time $t = 0$ its velocity is zero and its location is also zero. Determine its position at time $t = 4$ seconds.

Answer: $\dfrac{3008}{135} \approx 22.28$ ft

5.3 Elementary Area Computations
Solve the problem.

1)
Write the sum in expanded notation. $\sum\limits_{i=1}^{4} 2^i + 1$

Answer: $4 + 8 + 16 + 32$

2) Write the sum in expanded notation. $\sum_{k=1}^{10} \frac{1}{k}$

Answer: $1 + \frac{1}{2} + \frac{1}{3} + \frac{1}{4} + \frac{1}{5} + \frac{1}{6} + \frac{1}{7} + \frac{1}{8} + \frac{1}{9} + \frac{1}{10}$

3) Write the sum in expanded notation. $\sum_{n=1}^{5} (-1)^{n-1} x^{2n+1}$

Answer: $x^3 - x^5 + x^7 - x^9 + x^{11}$

4) Write the sum in summation notation. $1 + 3 + 7 + 15 + 33 + 63$

Answer: $\sum_{n=1}^{6} (2^n - 1)$

5) Write the sum in summation notation.

$\frac{1}{\sqrt{3}} + \frac{1}{\sqrt{8}} + \frac{1}{\sqrt{15}} + \frac{1}{\sqrt{24}} + \frac{1}{\sqrt{35}} + \frac{1}{\sqrt{48}} + \frac{1}{\sqrt{63}} + \frac{1}{\sqrt{80}} + \frac{1}{\sqrt{99}}$

Answer: $\sum_{x=2}^{10} \frac{1}{\sqrt{x^2-1}}$

6) Given $f(x) = x + 3$, find the exact area A of the region under $y = f(x)$ on the interval $[0, 3]$ by first computing $\sum_{i=1}^{n} f(x_i) \Delta x$ and then taking the limit as $n \to \infty$. (Hint: $\sum_{i=1}^{n} i = \frac{n(n+1)}{2}$)

Answer: $A = \frac{27}{2}$

7) Given $f(x) = 9 - 4x$, find the exact area A of the region under $y = f(x)$ on the interval $[1, 2]$ by first computing $\sum_{i=1}^{n} f(x_i) \Delta x$ and then taking the limit as $n \to \infty$. (Hint: $\sum_{i=1}^{n} i = \frac{n(n+1)}{2}$)

Answer: $A = 3$

8) Given $f(x) = x^2 + 3$, find the exact area A of the region under $y = f(x)$ on the interval $[1, 3]$ by first computing $\sum_{i=1}^{n} f(x_i) \Delta x$ and then taking the limit as $n \to \infty$. (Hint: $\sum_{i=1}^{n} i^2 = \frac{n(n+1)(2n+1)}{6}$)

Answer: $A = \frac{44}{3}$

5.4 Riemann Sums and the Integral
Solve the problem.

1) Express the given limit as a definite integral over the indicated interval.

$$\lim_{n \to \infty} \sum_{i=1}^{n} (3x_i + 4)\Delta x \qquad [2, 3]$$

Answer: $\int_{2}^{3} (3x + 4)dx$

2) Express the given limit as a definite integral over the indicated interval.

$$\lim_{n \to \infty} \sum_{i=1}^{n} (x_i^2 + 2x_i + 1)\Delta x \qquad [-1, 1]$$

Answer: $\int_{-1}^{1} (x^2 + 2x + 1)dx$

3) Express the given limit as a definite integral over the indicated interval.

$$\lim_{n \to \infty} \sum_{i=1}^{n} (\sqrt{32 - 2x_i^2})\Delta x \qquad [0, 4]$$

Answer: $\int_{0}^{4} (\sqrt{32 - 2x^2})dx$

4) Express the given limit as a definite integral over the indicated interval.

$$\lim_{n \to \infty} \sum_{i=1}^{n} (\cos 4\pi x_i)\Delta x \qquad [0, 2\pi]$$

Answer: $\int_{0}^{2\pi} \cos 4\pi x\, dx$

5) Given $f(x) = x^2 + x + 1$ on $[0, 4]$, with $n = 5$. Compute the Riemann sum $\sum_{i=1}^{n} f(x_i^*)\Delta x$ for f. (a) Use $x_i^* = x_{i-1}$, the left-hand endpoint of $[x_{i-1}, x_i]$. (b) Use $x_i^* = x_i$, the right-hand endpoint. (c) Use $x_i^* = m_i = \dfrac{(x_{i-1} + x_i)}{2}$, the midpoint of the ith subinterval.

Answer: (a) 21; (b) 41; (c) 30.75

6) Given $f(x) = x^3 - 1$ on $[1, 5]$, with $n = 4$. Compute the Riemann sum $\sum_{i=1}^{n} f(x_i^*)\Delta x$ for f. (a) Use $x_i^* = x_{i-1}$, the left-hand endpoint of $[x_{i-1}, x_i]$. (b) Use $x_i^* = x_i$, the right-hand endpoint. (c) Use $x_i^* = m_i = \dfrac{(x_{i-1} + x_i)}{2}$, the midpoint of the ith subinterval.

Answer: (a) 99; (b) 140; (c) 116.625

7) Given $f(x) = \sin x$ on $[0, \frac{\pi}{2}]$, with $n = 2$. Compute the Riemann sum

$\sum_{i=1}^{n} f(x_i^*)\Delta x$ for f. (a) Use $x_i^* = x_{i-1}$, the left-hand endpoint of $[x_{i-1}, x_i]$. (b) Use $x_i^* = x_i$, the right-hand endpoint. (c) Use $x_i^* = m_i = \frac{(x_{i-1} + x_i)}{2}$, the midpoint of the ith subinterval.

Answer: (a) ≈ 0.555; (b) ≈ 1.34; (c) ≈ 1.025

8) Evaluate $\int_0^6 (x^3 + 2x)dx$ by computing $\lim_{n \to \infty} \sum_{i=1}^{n} f(x_i)\Delta x$.

Answer: 360

9) Evaluate $\int_1^4 (6x - 3)dx$ by computing $\lim_{n \to \infty} \sum_{i=1}^{n} f(x_i)\Delta x$.

Answer: 36

10) Evaluate $\int_0^5 (2x^3 - 4x^2 + x + 1)dx$ by computing $\lim_{n \to \infty} \sum_{i=1}^{n} f(x_i)\Delta x$.

Answer: $\frac{490}{3} \approx 163.3$

11) Write and evaluate one Riemann sum for $f(x) = x^2$ over the interval $[1, 2]$ using a partition of $n = 5$ equal-length subintervals.

Answer: with x_i^* = left-endpoint

Riemann Sum = $\sum_{i=0}^{4} \frac{1}{5}\left(\frac{5+i}{5}\right)^2 = \frac{51}{25} = 2.04$

12) Evaluate $\lim_{n \to \infty} \frac{\sqrt{1} + \sqrt{2} + \sqrt{3} + \dots + \sqrt{n}}{n\sqrt{n}}$ by interpreting it as the limit of Riemann sums for some continuous function f defined on $[0, 1]$.

Answer: $f(x) = \sqrt{x}$ and limit $= \frac{2}{3} = \int_0^1 \sqrt{x}\, dx$

13)

Evaluate $\lim_{n \to \infty} \dfrac{\sin\frac{\pi}{n} + \sin\frac{2\pi}{n} + \sin\frac{3\pi}{n} + \ldots + \sin\frac{n\pi}{n}}{n}$ by interpreting it as the limit of Riemann sums for some continuous function f defined on $[0, 1]$.

Answer: $f(x) = \sin(\pi x)$ and limit $= \dfrac{2}{\pi} = \int_0^1 \sin(\pi x)\, dx$

14)

Use the definition of the definite integral to evaluate $\int_1^3 2x\, dx$. (Note: $\sum_{i=1}^{n} i = \dfrac{n(n+1)}{2}$.)

Answer: 8

5.5 Evaluation of Integrals
Solve the problem.

1)

Evaluate: $\int_3^5 x\sqrt{x^2 - 9}\, dx$

Answer: $\dfrac{64}{3}$

2)

Evaluate: $\int_0^{\pi} (\sin 4x)^6 \cos 4x\, dx$

Answer: 0

3)

Evaluate: $\int_{-1}^{1} (x^2 + 1)^4\, dx$

Answer: $\dfrac{2656}{315}$

4)

Evaluate: $\int_1^3 \dfrac{6}{x^2}\, dx$

Answer: 4

5)

Evaluate: $\displaystyle\int_0^{\pi/2} \cos x \, dx$

Answer: 1

6)

Evaluate: $\displaystyle\int_{-1}^{1} \sqrt{x+1} \, dx$

Answer: $\dfrac{4\sqrt{2}}{3}$

7)

Evaluate: $\displaystyle\int_{-2}^{-1} \dfrac{1}{x^2} \, dx$

Answer: $\dfrac{1}{2}$

8)

Evaluate: $\displaystyle\int_0^{1} (x^3 - x^2) \, dx$

Answer: $-\dfrac{1}{12}$

9)

Evaluate: $\displaystyle\int_1^{4} \dfrac{1}{\sqrt{x}} \, dx$

Answer: 2

10)

Evaluate: $\displaystyle\int_0^{\pi} \sin \dfrac{x}{4} \, dx$

Answer: $4 - 2\sqrt{2}$

11)

Evaluate: $\displaystyle\int_0^{1} \sqrt{x^4 + x^7} \, dx$

Answer: $\dfrac{4\sqrt{2} - 2}{9}$

12) Evaluate: $\int_0^\pi \cos x \, dx$

Answer: 0

13) Evaluate $\lim_{n \to \infty} \sum_{i=1}^{n} (\frac{1}{n^2} - 1)\frac{1}{n}$ by first recognizing the sum as a Riemann sum over a partition of [0, 1] and then evaluating the corresponding integral.

Answer: 0

14) Evaluate $\lim_{n \to \infty} \frac{1 + 2 + 3 + \ldots + n}{n^2 + n}$ by first recognizing the sum as a Riemann sum over a partition of [0, 1] and then evaluating the corresponding integral.

Answer: $\frac{1}{2}$

15) Evaluate $\lim_{n \to \infty} \frac{\sqrt[3]{1} + \sqrt[3]{2} + \sqrt[3]{3} + \ldots + \sqrt[3]{n}}{n\sqrt[3]{n}}$ by first recognizing the sum as a Riemann sum over a partition of [0, 1] and then evaluating the corresponding integral.

Answer: $\frac{3}{4}$

16) Evaluate $\lim_{n \to \infty} \sum_{i=1}^{n} \sec \pi i \, \frac{1}{n}$ by first recognizing the sum as a Riemann sum over a partition of [0, 1] and then evaluating the corresponding integral.

Answer: 0

17)

Given $\int_{-1}^{5} |3 - x| \, dx$, first sketch the graph $y = f(x)$ on the given interval. Then find the integral of f using your knowledge of area formulas for rectangles, triangles, and circles.

Answer: 10

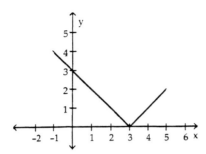

18)

Given $\int_{-4}^{3} (4 + |x|) \, dx$, first sketch the graph $y = f(x)$ on the given interval. Then find the integral of f using your knowledge of area formulas for rectangles, triangles, and circles.

Answer: $\dfrac{81}{2}$

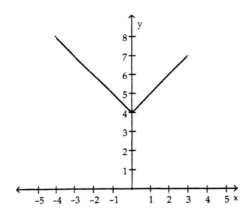

Calculus 125

19) Given $\int_{-2}^{4} (|3 - |3x||) \, dx$, first sketch the graph $y = f(x)$ on the given interval. Then find the integral of f using your knowledge of area formulas for rectangles, triangles, and circles.

Answer: 18

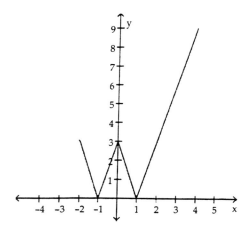

20) Given $\int_{0}^{6} (\sqrt{36 - x^2}) \, dx$, first sketch the graph $y = f(x)$ on the given interval. Then find the integral of f using your knowledge of area formulas for rectangles, triangles, and circles.

Answer: 9π

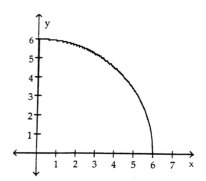

21) Find an upper and lower bound for the integral $\int_{0}^{1} \dfrac{1}{x+2} \, dx$ using the comparison properties of integrals.

Answer: $\dfrac{1}{3} \leq \int_{0}^{1} \dfrac{1}{x+2} \, dx \leq \dfrac{1}{2}$

22) Find an upper and lower bound for the integral $\int_0^{\pi/3} \sin^2 x\, dx$ using the comparison properties of integrals.

Answer: $\dfrac{\pi}{12} \leq \int_0^{\pi/3} \sin^2 x\, dx \leq \dfrac{\pi}{3}$

23) Suppose that a tank initially contains 2000 gal of water and the rate of change of its volume after the tank drains for t min is $V'(t) = (0.5)t - 30$ (in gallons per minute). How much water does the tank contain after it has been draining for 25 minutes?

Answer: ≈ 1406.25 gal

24) Suppose that the population of Cheyenne in 1985 was 55,000 and that its rate of growth t years later was $P'(t) = 1 + (0.5)t + (0.03)t^2$ (in thousands per year). What was its population in 2000?

Answer: $\approx 160,000$

5.6 The Fundamental Theorem of Calculus
Solve the problem.

1) Find the average value of $f(x) = x^3$ on $[0, 4]$.

Answer: 16

2) Find the average value of $g(x) = \sin x$ on $[0, \pi]$.

Answer: $\dfrac{2}{\pi}$

3) Find the average value of $h(x) = \sec^2 x$ on $[0, \dfrac{\pi}{4}]$.

Answer: $\dfrac{4}{\pi}$

4) Find the average value of $f(x) = \dfrac{1}{x^2}$ on $[1, 2]$.

Answer: $\dfrac{1}{2}$

5) Find the average value of $f(x) = \dfrac{1}{x^2}$ on $[1, 100]$.

Answer: $\dfrac{1}{100}$

6) Find the average value of $g(x) = \sqrt{x}$ on $[0, 4]$.

Answer: $\dfrac{4}{3}$

7) Find the average value of $h(x) = \cos x$ on $[0, 2\pi]$.

Answer: 0

8) Find the average value of $f(x) = x(x^2 + 1)^3$ on $[-1, 1]$.

Answer: 0

9) Evaluate: $\displaystyle\int_{-2}^{4} x\, dx$

Answer: 6

10) Evaluate: $\displaystyle\int_{-1}^{1} (x^2 + 4)^2\, dx$

Answer: $\dfrac{566}{15} \approx 37.7333$

11) Evaluate: $\displaystyle\int_{0}^{\pi/2} \cos x \sin x\, dx$

Answer: $\dfrac{1}{2}$

12) Evaluate: $\displaystyle\int_{-2}^{3} |x+1|\, dx$

Answer: $\dfrac{17}{2} = 8.5$

13) Evaluate: $\displaystyle\int_{0}^{4} |x^2 - \sqrt{x}|\, dx$

Answer: $\dfrac{50}{3}$

14) Evaluate: $\displaystyle\int_{0}^{\pi/6} \cos 2x\, dx$

Answer: $\dfrac{\sqrt{3}}{4}$

15) Suppose that a 1000-L water tank takes 20 min to drain and that after t minutes, the amount of water remaining in the tank is $V(t) = \frac{5}{2}(20-t^2)$ liters. What is the average amount of water in the tank during the time it drains.

Answer: $\frac{1000}{3} \approx 333$ L

16) On a certain day the temperature t hours past midnight was $T(t) = 60 + 5\sin(\frac{\pi}{6}(t - 10))$. What was the average temperature between noon and midnight?

Answer: $60°$

17) A sports car starts from rest ($x = 0$, $t = 0$) and experiences constant acceleration $x''(t) = a + 2t$ for T seconds. Find, in terms of a and T, (a) its final and average velocities and (b) its final and average positions.

Answer: (a) final velocity: $aT + T^2$; average velocity: $a + T$

(b) final position: $\frac{aT^2}{2} + \frac{T^3}{3}$; average position: $\frac{aT}{2} + \frac{T^2}{3}$

18) Apply the fundamental theorem of calculus to find the derivative of:

$$f(x) = \int_{-1}^{x^2} (t^2 + 1)^2 \, dt$$

Answer: $f'(x) = 2x(x+1)^2$

19) Apply the fundamental theorem of calculus to find the derivative of:

$$h(z) = \int_{2}^{z^2} \sqrt{u - 1} \, dz$$

Answer: $h'(z) = 2z\sqrt{z^2 - 1}$

20) Given $f(t) = \sqrt{t^2 + 4}$ on $[0, x^3]$. Apply the fundamental theorem of calculus to find $G'(x)$ if $G(x) = \int f(t)$.

Answer: $G'(x) = 3x^2\sqrt{x^6 + 4}$

21) Given $f(t) = \sin^2 t$ on $[0, x^2]$. Apply the fundamental theorem of calculus to find $G'(x)$ if $G(x) = \int f(t)$.

Answer: $G'(x) = 2x\sin^2 x^2$

22) Given $f(x) = \int_{2}^{4x} \cos t^3 \, dt$, find $f'(x)$.

Answer: $f'(x) = 4\cos 64x^3$

23)

Given $f(x) = \int_{2}^{(x^3 + x)} \frac{dt}{t}$, find $f'(x)$.

Answer: $f'(x) = \dfrac{3x^2 + 1}{x^3 + x}$

5.7 Integration by Substitution
Solve the problem.

1) Solve using the indicated substitution. $\int x^3(1 + x^4)^{3/2}\, dx;\ u = x^4$

   Answer: $\dfrac{(1 + x^4)^{5/2}}{10} + C$

2) Solve using the indicated substitution. $\int x^3(1 + x^4)^{3/2}\, dx;\ u = 1 + x^4$

   Answer: $\dfrac{(1 + x^4)^{5/2}}{10} + C$

3) Solve using the indicated substitution. $\int \dfrac{\sin x}{(1 + \cos x)^7}\, dx;\ u = \cos x$

   Answer: $\dfrac{1}{6(1 + \cos x)^6} + C$

4) Solve using the indicated substitution. $\int \dfrac{\sin x}{(1 + \cos x)^7}\, dx;\ u = 1 + \cos x$

   Answer: $\dfrac{1}{6(1 + \cos x)^6} + C$

5) Solve using the indicated substitution. $\int (\sec 6x)^2\, dx;\ u = 6x$

   Answer: $\dfrac{1}{6} \tan 6x + C$

6) Solve using the indicated substitution. $\int \sec^7 x \tan x\, dx;\ u = \sec^6 x$

   Answer: $\dfrac{1}{7} \sec^7 x + C$

7) Solve using the indicated substitution. $\int x\sqrt{x + 1}\, dx;\ u = x + 1$

   Answer: $\dfrac{2(x + 1)^{3/2}(3x - 2)}{15} + C$

8) Solve using the indicated substitution. $\int \cos 8x\, dx;\ u = 8x$

   Answer: $\dfrac{1}{8} \sin 8x + C$

9) Solve using the indicated substitution. $\int_0^1 \frac{x^3}{\sqrt{1+x^4}} dx; u = 1 + x^4$

Answer: $\frac{\sqrt{2}-1}{2}$

10) Solve using the indicated substitution. $\int_0^1 x^3\sqrt{1+x^4} \, dx; u = 1 + x^4$

Answer: $\frac{2\sqrt{2}-1}{6}$

11) Solve using the indicated substitution. $\int_0^\pi \sin^2 x \cos x \, dx; u = \sin x$

Answer: 0

12) Solve using the indicated substitution. $\int_0^\pi (1 + \sin x)^7 \cos x \, dx; u = \sin x$

Answer: 0

13) Solve using the indicated substitution. $\int_{-1}^2 x^2\sqrt{1+x^3} \, dx; u = 1 + x^3$

Answer: 6

14) Evaluate: $\int \frac{dx}{\sqrt{4+5x}}$

Answer: $\frac{2\sqrt{4+5x}}{5} + C$

15) Evaluate: $\int \cos(2\pi x + 2) \, dx$

Answer: $\frac{\sin(2\pi x + 2)}{2\pi} + C$

16) Evaluate: $\int x^3 \sin(4x^4) \, dx$

Answer: $\frac{-\cos 4x^4}{16} + C$

17) Evaluate: $\int \frac{\sin\sqrt{2x}}{\sqrt{2x}} dx$

Answer: $-\cos\sqrt{2x} + C$

18) Evaluate: $\int_{1}^{3} \frac{dt}{(t+1)^2}$

Answer: $\frac{1}{4}$

19) Evaluate: $\int_{0}^{2} x^2\sqrt{x^3+1}\, dx$

Answer: $\frac{52}{9}$

20) Evaluate: $\int_{1}^{4} \frac{(4+\sqrt{x})^2}{2\sqrt{x}}\, dx$

Answer: $\frac{91}{3}$

21) Evaluate: $\int_{0}^{\sqrt{\pi/2}} t\sin t^2\, dt$

Answer: $\frac{2-\sqrt{2}}{4}$

22) Evaluate: $\int_{0}^{\pi^2/2} \frac{2\sin\sqrt{2x}\cos\sqrt{2x}}{3\sqrt{2x}}\, dx$

Answer: 0

23) Evaluate: $\int \frac{\sin^2 x}{2}\, dx$ (hint: $\sin^2\theta = \frac{1-\cos 2\theta}{2}$)

Answer: $\frac{x}{4} - \frac{\sin 2x}{8} + C$

24) Evaluate: $\int 2\cos^2 x\, dx$ (hint: $\cos^2\theta = \frac{1+\cos 2\theta}{2}$)

Answer: $x + \frac{\sin 2x}{2} + C$

25) Evaluate: $\int_{0}^{\pi/2} 4\sin^2 2t\, dt$ (hint: $\sin^2\theta = \frac{1-\cos 2\theta}{2}$)

Answer: π

26) Evaluate: $\int_0^1 \cos^2 4\pi x \, dx$ (hint: $\cos^2\theta = \dfrac{1+\cos 2\theta}{2}$)

Answer: $\dfrac{1}{2}$

27) Evaluate: $\int (\tan^2 x + \sin x) \, dx$ (hint: $\tan^2\theta + 1 = \sec^2\theta$)

Answer: $\tan x - \cos x - x + C$

28) Evaluate: $\int_0^{\pi/4} (x + \tan^2 x) \, dx$ (hint: $\tan^2\theta + 1 = \sec^2\theta$)

Answer: $\dfrac{\pi^2 - 8\pi + 32}{32} = \dfrac{\pi^2}{32} - \dfrac{\pi}{4} + 1$

5.8 Areas of Plane Regions
Solve the problem.

1) Sketch the region bounded by the graphs of: $f(x) = x^2$, $g(x) = x^3$ and then find its area.

Answer: $\dfrac{1}{12}$

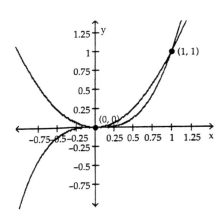

2) Sketch the region bounded by the graphs of : $f(x) = x^3$, $g(x) = x^4$ and then find its area.

Answer: $\dfrac{1}{20}$

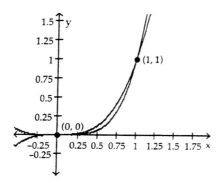

3) Sketch the region bounded by the graphs of : $f(x) = x^2$, $g(x) = 8 - x^2$ and then find its area.

Answer: $\dfrac{64}{3}$

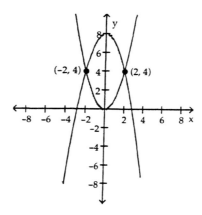

4) Sketch the region bounded by the graphs of : $x = y^2$, $x = 18 - y^2$ and then find its area.

   Answer: 72

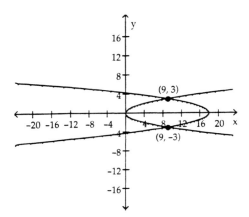

5) Sketch the region bounded by the graphs of : $f(x) = x^4$, $y = 1$ and then find its area.

   Answer: $\dfrac{8}{5}$

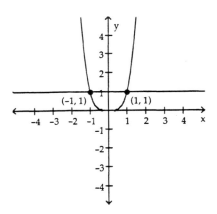

6) Sketch the region bounded by the graphs of : $x = y^2$, $y = x^2$ and then find its area.

Answer: $\dfrac{1}{3}$

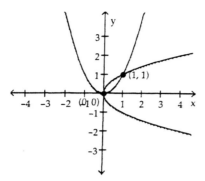

7) Sketch the region bounded by the graphs of : $f(x) = x$, $f(x) = 2x$, $f(x) = 3 - x$ and then find its area.

Answer: $\dfrac{3}{4}$

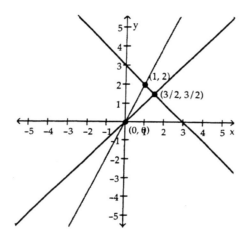

8) Sketch the region bounded by the graphs of : $f(x) = x^4$, $y = 0$, $x = 1$ and then find its area.

Answer: $\dfrac{1}{5}$

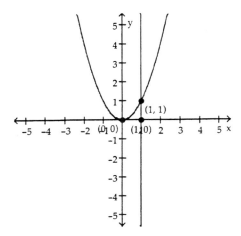

9) Sketch the region bounded by the graphs of : $x = y - y^3$, $x = 0$ and then find its area.

Answer: $\dfrac{1}{4}$ each $= \dfrac{1}{2}$

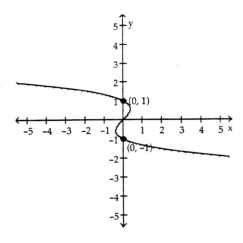

Calculus 137

10) Sketch the region bounded by the graphs of : $y = x\sqrt{1-x^2}$, $y = 0$ and then find its area.

Answer: $\dfrac{1}{3}$ each = $\dfrac{2}{3}$

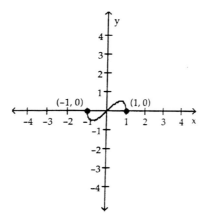

11) Sketch the region bounded by the graphs of : $y^2 = 4x$, $y = 2x - 4$ and then find its area.

Answer: 9

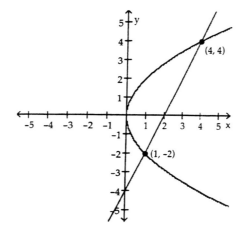

12) Sketch the region bounded by the graphs of : $x = 6 - y^2$, $x = y$ and then find its area.

Answer: $\dfrac{125}{6}$

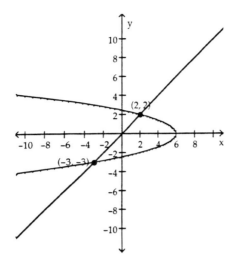

13) Sketch the region bounded by the graphs of : $y = 9 - x^2$, $y = 5 - 3x$ and then find its area.

Answer: $\dfrac{125}{6}$

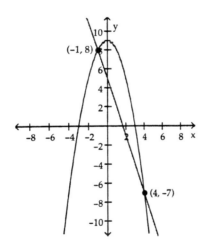

14) Find the area of the region bounded by the graphs of : $f(x) = x^n$, $g(x) = x^{n+1}$ (n is a fixed positive integer)

Answer: $\dfrac{1}{(n+1)(n+2)}$

15) Find a number $k > 0$ such that the area bounded by the curves $y = 2x^2$ and $y = 3k - x^2$ is 32.

Answer: $k = 4$

16) Find a number $k > 0$ such that the line $y = k$ divides the region between the parabola $y = 10 - x^2$ and the x-axis into two regions having equal areas.

Answer: $k = 3$

5.9 Numerical Integration
Solve the problem.

1) Given $\int_0^4 2x\, dx$ with $n = 4$, calculate (a) the trapezoidal approximation T_n, (b) the midpoint approximation M_n, and (c) the exact value of the integral.

Answer: (a) 16; (b) 16; (c) 16

2) Given $\int_0^3 x^2\, dx$ with $n = 4$, calculate (a) the trapezoidal approximation T_n, (b) the midpoint approximation M_n, and (c) the exact value of the integral.

Answer: (a) 9.28125; (b) 8.859; (c) 9

3) Given $\int_0^4 \sqrt{x}\, dx$ with $n = 2$, calculate (a) the trapezoidal approximation T_n, (b) the midpoint approximation M_n, and (c) the exact value of the integral.

Answer: (a) 3.621; (b) 4.098; (c) $\dfrac{16}{3}$

4) Given $\int_1^5 \dfrac{1}{2x^2}\, dx$ with $n = 8$, calculate (a) the trapezoidal approximation T_n, (b) the midpoint approximation M_n, and (c) the exact value of the integral.

Answer: (a) 0.4198; (b) 0.39035; (c) 0.4

5) Given $\int_0^\pi \cos x\, dx$ with $n = 4$, calculate (a) the trapezoidal approximation T_n, (b) the midpoint approximation M_n, and (c) the exact value of the integral.

Answer: (a) 0; (b) 1.5; (c) 0

6) Given $\int_0^{\pi/2} \sin x\, dx$ with $n = 2$, calculate (a) the trapezoidal approximation T_n, (b) the midpoint approximation M_n, and (c) the exact value of the integral.

Answer: (a) 0.948; (b) 1.026; (c) 1

7) Approximate the integral $\int_0^1 \sqrt{1+x^3}\, dx$; $n = 4$, by (a) first applying Simpson's Rule and (b) then applying the trapezoidal rule.

Answer: (a) 1.1114; (b) 1.1314

8) Approximate the integral $\int_1^2 \dfrac{1}{x^2}\, dx$; $n = 6$, by (a) first applying Simpson's Rule and (b) then applying the trapezoidal rule.

Answer: (a) 0.50009; (b) 0.50402

9) Approximate the integral $\int_0^1 x^3\, dx$; $n = 4$, by (a) first applying Simpson's Rule and (b) then applying the trapezoidal rule.

Answer: (a) 0.25; (b) 0.265625

10) Approximate the integral $\int_0^1 x^4\, dx$; $n = 4$, by (a) first applying Simpson's Rule and (b) then applying the trapezoidal rule.

Answer: (a) 0.200521; (b) 0.2207

11) Approximate the integral $\int_0^1 x^4\, dx$; $n = 6$, by (a) first applying Simpson's Rule and (b) then applying the trapezoidal rule.

Answer: (a) 0.200102; (b) 0.19985

12) Approximate the integral $\int_0^\pi \dfrac{\sin x}{x}\, dx$; $n = 10$, by (a) first applying Simpson's Rule and (b) then applying the trapezoidal rule.

Answer: (a) 1.85184; (b) 1.8493

13) Calculate (a) the trapezoidal approximation and (b) Simpson's approximation to $\int_a^b f(x)\, dx$ where f is the tabulated function.

x	$a=1.0$	1.33	1.67	2.0	2.33	2.67	$3.0=b$
$f(x)$	5.2	6.9	1.4	0.06	2.3	0.01	1.5

Answer: (a) 4.46733; (b) 4.6644

14) Calculate (a) the trapezoidal approximation and (b) Simpson's approximation to $\int_a^b f(x)\, dx$ where f is the tabulated function.

x	$a=0$	1	2	3	4	5	6	7	8	9	$10=b$
$f(x)$	1	2	3	4	5	6	7	8	9	10	11

Answer: (a) 60; (b) 60

15) The graph shows the measured rate of water flow (in liters per minute) into a tank during a 10 minute period. Using 10 subintervals in each case, estimate the total amount of water that flows into the tank during this period by using (a) the trapezoidal approximation and (b) Simpson's approximation.

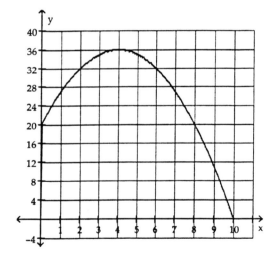

Answer: (a) 262 L; (b) 262.67 L

16) The graph below shows the daily mean temperature recorded during September at Estes Park, Colorado. Using 10 subintervals in each case, estimate the average temperature during that month by using (a) the trapezoidal approximation and (b) Simpson's approximation.

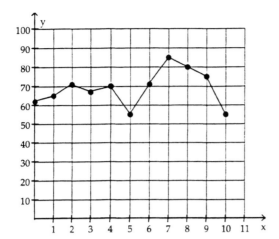

Answer: (a) 69.4°; (b) 69.2°

17) Use Simpson's approximation with $n = 2$ to calculate (without implicit integration) the area of the region between the graphs of $y = x^2 - 5x$ and $y = 6 - 4x$.

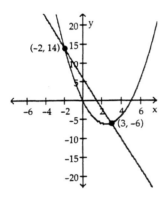

Answer: $\dfrac{125}{6}$

Ch. 6 Applications of the Integral
6.1 Riemann Sum Approximations
Solve the problem.

1) Evaluate $\lim\limits_{n \to \infty} \sum\limits_{i=1}^{n} 4x_i^* \, \Delta x$ with $a = 0$, $b = 3$, by computing the value of the appropriate related integral.

   Answer: 18

2) Evaluate $\lim\limits_{n \to \infty} \sum\limits_{i=1}^{n} \cos\pi x_i^* \, \Delta x$ with $a = 0$, $b = \frac{1}{2}$, by computing the value of the appropriate related integral.

   Answer: $\frac{1}{\pi}$

3) Evaluate $\lim\limits_{n \to \infty} \sum\limits_{i=1}^{n} [2(x_i^*)^2 - 3] \, \Delta x$ with $a = -1$, $b = 2$, by computing the value of the appropriate related integral.

   Answer: -3

4) Evaluate $\lim\limits_{n \to \infty} \sum\limits_{i=1}^{n} \frac{1}{(x_i)^2} \, \Delta x$ with $a = 2$, $b = 4$, by computing the value of the appropriate related integral.

   Answer: $\frac{1}{4}$

5) Express $\lim\limits_{n \to \infty} \sum\limits_{i=1}^{n} \frac{3\pi}{2} x_i \, f(x_i) \Delta x$ with $a = 0$, $b = 1$, as an integral involving the function f.

   Answer: $\int\limits_{0}^{1} \frac{3\pi}{2} x f(x) \, dx$

6) Express $\lim\limits_{n \to \infty} \sum\limits_{i=1}^{n} 2[f(x_i)]^3 \, \Delta x$ with $a = 1$, $b = 4$, as an integral involving the function f.

   Answer: $\int\limits_{1}^{4} 2[f(x)]^3 \, dx$

7) Express $\lim_{n \to \infty} \sum_{i=1}^{n} \sqrt{6 + [f(x_i)]^3} \, \Delta x$ with $a = 2$, $b = 9$, as an integral involving the function f.

Answer: $\int_{2}^{9} \sqrt{6 + [f(x)]^3} \, dx$

8) Express $\lim_{n \to \infty} \sum_{i=1}^{n} 4\pi m_i \sqrt[3]{[f(m_i)]^3 - 2} \, \Delta x$ with $a = 2$, $b = 8$, as an integral involving the function f.

Answer: $\int_{2}^{8} 4\pi x \sqrt[3]{[f(x)]^3 - 2} \, dx$

9) Find the mass M (in grams) of a rod coinciding with the interval [0, 21] which has the density function $\rho(x) = \frac{x^2}{2}$.

Answer: 1543.5 grams

10) Find the mass M (in grams) of a rod coinciding with the interval [0, 15] which has the density function $\rho(x) = 100 - 3x$.

Answer: 1162.5 grams

11) Find the mass M (in grams) of a rod coinciding with the interval [0, 50] which has the density function $\rho(x) = 3x(x - 12)$.

Answer: 80,000 grams or 80 kg

12) Find the mass M (in grams) of a rod coinciding with the interval [0, 4] which has the density function $\rho(x) = 5\sin\frac{\pi}{4}x$.

Answer: $\frac{40}{\pi}$ grams $\approx$ 12.7324 grams

13) A particle moving along a line has the velocity function $v = f(t) = 3t + 4$ between time $t = a = 1$ and $t = b = 6$. Compute (a) the net distance and (b) the total distance travelled.

Answer: (a) 75.5; (b) 75.5

14) A particle moving along a line has the velocity function $v = f(t) = 6t - 18$ between time $t = a = 0$ and $t = b = 5$. Compute (a) the net distance and (b) the total distance travelled.

Answer: (a) −15; (b) 39

15) A particle moving along a line has the velocity function $v = f(t) = \sin 2t$ between time $t = a = 0$ and $t = b = \pi$. Compute (a) the net distance and (b) the total distance travelled.

Answer: (a) 0; (b) 2

16) A particle moving along a line has the velocity function $v = f(t) = \cos\pi t$ between time $t = a = -\frac{1}{4}$ and $t = b = 1$.

Compute (a) the net distance and (b) the total distance travelled.

Answer: (a) $\frac{\sqrt{2}}{2\pi}$; (b) $\frac{2}{\pi} + \frac{\pi\sqrt{2}}{2}$

17) A particle moving along a line has the velocity function $v = f(t) = t^2 - 8t + 12$ between time $t = a = 1$ and $t = b = 7$. Compute (a) the net distance and (b) the total distance travelled.

Answer: (a) -6; (b) $\frac{46}{3} \approx 15.33$

18) Suppose that the rate of water flow into an initially empty swimming pool is $500 - 6t$ gallons per minute at time t (in minutes). How much water flows into the pool during the interval $t = 15$ to $t = 30$?

Answer: 5475 gallons

19) Suppose that the birth rate in Ft. Collins t years after 1968 was $12 + 5t$ thousands per year and the death rate t years after 1968 was $4 + \frac{1}{3}t$ thousands per year. Set up and evaluate an appropriate integral to compute (a) the total number of births between 1968 and 1980 and (b) the total number of deaths in that same time period. (c) The population of Ft. Collins in 1968 was 565,000. What was the population in 1980?

Answer: (a) 504,000; (b) 96,000; (c) 973,000

6.2 Volumes by the Method of Cross Sections
Solve the problem.

1) The plane region R is bounded above by the graph of $y = x^{-2}$ and below by the x-axis over the interval $[1, 3]$. Find the volume generated by rotating R around the x-axis.

Answer: $\frac{26\pi}{81}$

2) The region R is bounded by the graph of $y = x^3$, the y-axis, and the horizontal line $y = 1$. Find the volume generated by rotating R around the y-axis.

Answer: $\frac{3\pi}{5}$

3) The region R is bounded by the graph of $y = x^3$, the y-axis, and the horizontal line $y = 1$. Find the volume generated by rotating R around the x-axis.

Answer: $\frac{6\pi}{7}$

4) The plane region R is bounded above by the graph of $y = x - x^3$ and below by the x-axis over the interval $[0, 1]$. Find the volume generated by rotating R around the x-axis.

Answer: $\frac{8\pi}{105}$

5) The region R in the first quadrant is bounded by the coordinate axes and the graph of $y = 4 - x^2$. When it is rotated around the y-axis it generates a solid of volume V. Find V.

Answer: 8π

6) The region R in the first quadrant is bounded above by the graph of $y = \sqrt{\sin x}$ and below by the x-axis, $0 \leq x \leq \pi$. When it is rotated around the x-axis it generates a solid of volume V. Find V.

   Answer: 2π

7) Find the volume generated when the region bounded by the y-axis, the graph of $xy^2 = 1$, and the horizontal lines $y = 1$ and $y = 2$ is rotated around the y-axis.

   Answer: $\dfrac{21\pi}{8}$

8) Find the volume of the solid that is generated by rotating the region formed by the graphs of $y = x^2$, $y = 2$, and $x = 0$ about the y-axis.

   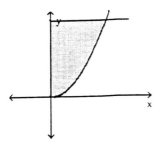

   Answer: 2π

9) Find the volume of the solid that is generated by rotating the region formed by the graphs of $y = x^2$, $y = 2$, and $x = 0$ about the x-axis.

   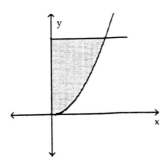

   Answer: $\dfrac{16\pi\sqrt{2}}{5}$

10) Find the volume of the solid that is generated by rotating the region formed by the graphs of $y = 2\sin 2x$, $y = 0$, and $x = 0$ over the interval $[0, \dfrac{\pi}{2}]$ about the x-axis.

    Answer: π^2

11) Find the volume of the solid that is generated by rotating the region formed by the graphs of $y = x^2$ and $y = 2x$ about the line $x = 2$.

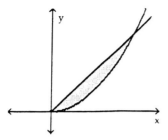

Answer: $\dfrac{8\pi}{3}$

12) Find the volume of the solid that is generated by rotating the region formed by the graphs of $y = 2x^2$ and $y = 4x$ about the line $x = 3$.

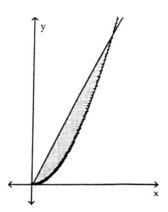

Answer: $\dfrac{32\pi}{3}$

13) Find the volume of the solid that is generated by rotating the region formed by the graphs of $y = 2x^2$ and $y = 4x$ about the line $y = 10$.

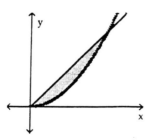

Answer: $\dfrac{544\pi}{15} \approx 36.27\pi$

14) Find the volume of the solid that is generated by rotating the region formed by the graphs of $y = x^2$ and $x = y^2$ about the line $y = -1$.

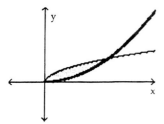

Answer: $\dfrac{19\pi}{30}$

15) Find the volume of the solid that is generated by rotating the region formed by the graphs of $y = x^2$ and $x = y^2$ about the line $x = -4$.

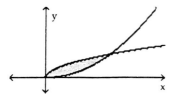

Answer: $\dfrac{89\pi}{30}$

16) Find the volume of the solid that is generated by rotating the region formed by the graphs of $y = 6 - x^2$ and $y = x^2 + 2x - 6$ about the line $y = -7$.

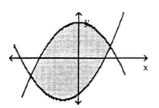

Answer: $\dfrac{1625\pi}{3}$

17) R is the region bounded by the curve $y = 2\cos x$ and the x-axis for $0 \le x \le \dfrac{\pi}{2}$. Find the volume of the solid obtained by revolving R around the x-axis.

Answer: π^2

18) R is the region between $x = -\dfrac{\pi}{6}$ and $x = \dfrac{\pi}{6}$ that is bounded by the curves $y = \cos x$ and $y = \dfrac{1}{3}$. Find the volume of the solid obtained by revolving R around the x-axis.

Answer: $\dfrac{27\pi\sqrt{3} + 14\pi^2}{108} = \dfrac{\pi\sqrt{3}}{4} + \dfrac{7\pi^2}{54} \approx 2.63974$

19) R is the region bounded by the curves $y = \tan x$, $y = \cot x$, and the x-axis for $0 \le x \le \frac{\pi}{2}$. Find the volume of the solid obtained by revolving R around the x-axis. (*hint*: you may have to divide R into 2 regions)

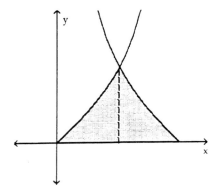

Answer: $2\pi - \frac{\pi^2}{2}$

20) The base of a certain solid is a circular disk with diameter AB of length a. Find the volume of the solid if each cross section perpendicular to AB is a rectangle with length = twice width.

Answer: $\frac{4}{3}a^3$

21) A solid has base B in the coordinate plane. The base B is the region between the x-axis and the graph of $y = x - x^2$ over the interval $[0, 1]$. Every cross section of the solid in a plane perpendicular to the x-axis is a square. Find the volume of the solid.

Answer: $\frac{1}{30}$

22) The base of a certain solid is a circular disk with diameter AB of length $2a$. Find the volume of the solid if each cross section perpendicular to AB is an isosceles right triangle.

Answer: $\frac{8}{3}a^3$

23) The region R is bounded by the graphs of $x - 2y = 3$ and $x = y^2$. Set up (but do not evaluate) the integral that gives the volume of the solid obtained by rotating R around the line $x = -1$.

Answer: $\int_{-1}^{3} \pi[(2y + 4)^2 - (y^2 + 1)^2]\,dy$

24) A solid has its base in the xy-plane; the base is the region bounded by $x = y^2$ and $x = 4$. Each vertical cross section of the region perpendicular to the x-axis is a semicircle with its diameter in the base. Find the volume of the solid.

Answer: 4π

25) Set up (but do not evaluate) the integral that gives the volume of the following solid: Its base is a circular disk of radius 1 in the xy-plane; cross sections perpendicular to a fixed diameter of the base are equilateral triangles.

Answer: $\int_{-1}^{1} \frac{x^2\sqrt{3}}{2} dx$

26) The graph of $x^2 + y^2 = R^2$ is rotated around the x-axis to form a sphere of radius R. Find the volume of the sphere by methods of integral calculus.

Answer: $\frac{4}{3}\pi R^3$

6.3 Volume by the Method of Cylindrical Shells
Solve the problem.

1) The region R is the first-quadrant region bounded by the coordinate axes and the graph of $y = 2 + 3x - x^3$, $0 \leq x \leq 2$. Find the volume generated by the rotation of R around the y-axis.

Answer: $\frac{56\pi}{5}$

2) The region R is bounded below by the x-axis and above by the graph of $y = 2x - x^2$. Find the volume generated by rotation of R around
(a) the y-axis; (b) the line $y = -1$.

Answer: (a) $\frac{8\pi}{3}$; (b) $\frac{16\pi}{3}$

3) The region R lies in the first quadrant and is bounded by the graph of $y = \frac{1}{(x^2+1)^2}$, the coordinate axes, and the vertical line $x = 2$. Find the volume generated when R is rotated around the y-axis.

Answer: $\frac{4\pi}{5}$

4) The plane region R is bounded above by the graph of $y = x - x^3$ and below by the x-axis over the interval $[0, 1]$. Find the volume generated when R is rotated around the y-axis.

Answer: $\frac{4\pi}{15}$

5) The plane region R is bounded above by the graph of $y = x - x^3$ and below by the x-axis over the interval of $[0, 1]$. Find the volume generated when R is rotated around the vertical line $x = 2$.

Answer: $\frac{11\pi}{15}$

6) The region R in the first quadrant is bounded by the coordinate axes and the graph of $y = 4 - x^2$. Fine the volume generated when R is rotated around the y-axis.

Answer: 8π

7) The plane region Q is bounded above by the graph of $x = 16 - y^2$, on the left by the y-axis, and below by the x-axis. Find the volume generated when Q is rotated around the x-axis.

Answer: 128π

8) Use the method of cylindrical shells to find the volume of the solid rotated about the y-axis given the conditions: $x = y^2$, $(y \geq 0)$; $y = 4 - \sqrt{x}$; $x = 1$.

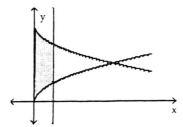

Answer: $\dfrac{12\pi}{5}$

9) Use the method of cylindrical shells to find the volume of the solid rotated about the line $y = 1$ given the conditions: $x = y^2 + 3y$; $y = 0$; $x = 0$.

Answer: 9π

10) Use the method of cylindrical shells to find the volume of the solid rotated about the line $x = -1$ given the conditions: $y = x^3 - x^2$; $y = 0$; $x = 0$.

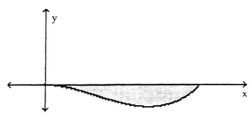

Answer: $\dfrac{4\pi}{15}$

11) Use the method of cylindrical shells to find the volume of the solid rotated about the line $x = -2$ given the conditions: $y = \dfrac{x^2}{4}$; $y = x$.

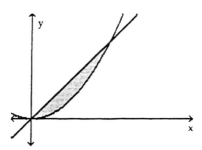

Answer: $\dfrac{64\pi}{3}$

12) Use the method of cylindrical shells to find the volume of the hyperboloid obtained by revolving the hyperbolic region bounded by the graph of the equation $(\frac{x}{a})^2 - (\frac{y}{b})^2 = 1$ around the x-axis.

Answer: $\frac{4ab^2\pi}{3}(2\sqrt{2} - 1)$

13) Find the volume of the solid generated by revolving the region bounded by the curves $y = x^2$ and $y = 2x + 3$ around (a) the line $x = 2$ and (b) the line $y = -1$.

Answer: (a) $\frac{17\pi}{3}$; (b) $\frac{448\pi}{5}$

6.4 Arc Length and Surface Area of Revolution
Solve the problem.

1) Set up (but do not evaluate) the integral that gives the length of the graph of $y = x^4$ from (0, 0) to (1, 1).

Answer: $\int_0^1 \sqrt{1 + 16x^6}\, dx$

2) Set up (but do not evaluate) the integral that gives the length of the graph of $y = \sin x$ from $x = 0$ to $x = \pi$.

Answer: $\int_0^\pi \sqrt{1 + \cos^2 x}\, dx$

3) Set up (but do not evaluate) the integral that gives the length of the graph of $y = 4x^3 - 3x^4$, with $0 \le x \le 4$.

Answer: $\int_0^4 \sqrt{1 + 144x^4(1 - x)^2}\, dx$

4) Set up (but do not evaluate) the integral that gives the length of the graph of $x = 4 - y^2$, with $0 \le y \le 2$.

Answer: $\int_0^2 \sqrt{1 + 4y^2}\, dy$

5) Set up (but do not evaluate) the integral that gives the length of the graph of $y = \cos x$, with $0 \le x \le \frac{\pi}{2}$.

Answer: $\int_0^{\pi/2} \sqrt{1 + \sin^2 x}\, dx$

6) Set up (but do not evaluate) the integral that gives the surface area of revolution generated by rotating the graph of $y = x^4$, $0 \le x \le 1$, around the y-axis.

Answer: $\int_0^1 2\pi x\sqrt{1 + 16x^6}\, dx$

Calculus 153

7) Set up (but do not evaluate) the integral that gives the surface area of revolution generated by rotating the graph of $y = x^2 + 1$, $0 \le x \le 2$, around the y-axis.

Answer: $\int_0^2 2\pi x \sqrt{1 + 4x^2}\, dx$

8) Set up (but do not evaluate) the integral that gives the surface area of revolution generated by rotating the graph of $y = x^2 + 1$, $0 \le x \le 2$, around the line $x = 1$.

Answer: $\int_0^2 2\pi(x - 1)\sqrt{1 + 4x^2}\, dx$

9) Set up (but do not evaluate) the integral that gives the surface area of revolution generated by rotating the graph of $y = x - x^2$, $0 \le x \le 1$, around the x-axis.

Answer: $\int_0^1 2\pi(x - x^2)\sqrt{1 + (1 - 2x)^2}\, dx$

10) Set up (but do not evaluate) the integral that gives the surface area of revolution generated by rotating the graph of $y = 4\sqrt{x + 1}$, $-1 \le x \le 0$, around the x-axis.

Answer: $\int_{-1}^{0} 8\pi (\sqrt{x + 1}) \sqrt{1 + \dfrac{4}{x + 1}}\, dx$

11) Find the length of the graph of $y = \dfrac{x^4}{16} + \dfrac{1}{2x^2}$, over the interval $1 \le x \le 2$.

Answer: $\dfrac{21}{16}$

12) Find the length of the graph of $y = \dfrac{1}{3}x^{3/2} - x^{1/2}$ from $(1, -\dfrac{2}{3})$ to $(4, \dfrac{2}{3})$.

Answer: $\dfrac{10}{3}$

13) Find the length of the graph of $y = \dfrac{2}{3}(x^2 - 1)^{3/2}$ from $x = 0$ to $x = 4$.

Answer: $\dfrac{116}{3}$

14) Find the length of the graph of $x = \dfrac{2}{3}(y + 1)^{3/2}$ from $y = 0$ to $y = 7$.

Answer: $18 - \dfrac{4\sqrt{2}}{3}$

15) Find the length of the graph of $12xy - 3y^4 = 4$ from $(\dfrac{7}{12}, 1)$ to $(\dfrac{13}{6}, 2)$.

Answer: $\dfrac{23}{12}$

16) Find the length of the graph of $y = \frac{2}{3}x^{3/2}$ from (0, 0) to (9, 18).

Answer: $\frac{2}{3}(10\sqrt{10} - 1)$

17) Find the area of the surface of revolution generated by rotating $y = x^3$, $0 \le x \le 1$ around the x-axis.

Answer: $\frac{\pi}{27}(10\sqrt{10} - 1)$

18) Find the area of the surface obtained when the graph of $y = x^2$, $0 \le x \le 1$, is rotated around the y-axis.

Answer: $\frac{\pi}{6}(5\sqrt{5} - 1)$

19) Find the area of the surface of revolution generated by revolving $x = \sqrt{y}$; $0 \le y \le 4$ around the y-axis.

Answer: $\frac{\pi}{6}(17\sqrt{17} - 1)$

20) Find the area of the surface of revolution generated by rotating $y = 4x^2$; $0 \le x \le 2$ around the y-axis.

Answer: $\frac{\pi}{96}(257\sqrt{257} - 1)$

21) Find the area of the surface of revolution generated by rotating $y^2 = 2x$; $0 \le x \le 8$ around the x-axis.

Answer: $\frac{2\pi}{3}(17\sqrt{17} - 1)$

22) Find (a) the total length of the astroid shown and (b) the total area generated by rotation of the astroid around the y-axis. Equation of astroid: $x^{2/3} + y^{2/3} = 4$

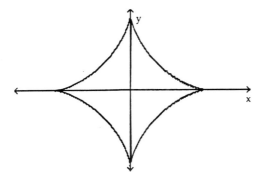

Answer: (a) 48; (b) $\frac{1536\pi}{5}$

6.5 Force and Work
Solve the problem.

1) Find the work done by the force $F(x) = 5$ in moving a particle along the x-axis from $a = -1$ to $b = 5$.

Answer: 30

2) Find the work done by the force $F(x) = 2x - 3$ in moving a particle along the x-axis from $a = 1$ to $b = 6$.

Answer: 20

3) Find the work done by the force $F(x) = \dfrac{5}{x^3}$ in moving a particle along the x-axis from $a = 1$ to $b = 5$.

   Answer: $\dfrac{12}{5} = 2.4$

4) Find the work done by the force $F(x) = \dfrac{-\sqrt{x}}{2}$ in moving a particle along the x-axis from $a = 0$ to $b = 9$.

   Answer: -9

5) Find the work done by the force $F(x) = \cos\dfrac{\pi}{2}x$ in moving a particle along the x-axis from $a = -1$ to $b = 1$.

   Answer: $\dfrac{4}{\pi}$

6) A 100-ft length of steel chain weighing 15 lb/ft is hanging from the top of a tall building. How much work is done in pulling all of the chain to the top of the building?

   Answer: 1500 ft · lb

7) A 10-ft trough filled with water has a semicircular cross section of diameter 4 ft. How much work is done in pumping all the water over the edge of the trough? Assume that water weighs 62.5 lb/ft^3.

   Answer: 2500π ft · lb

8) A cylindrical tank of radius R feet and length L feet is lying on its side on horizontal ground. Oil weighing δ pounds per cubic foot is available at ground level and is to be pumped into the tank. Find the work required to do so.

   Answer: $R^3 L\pi\delta$ ft · lb

9) The graph of $y = x^3$, $0 \le x \le 1$, is rotated around the y-axis to form the surface of a tank. If units on the coordinate axes are in feet, find how much work is done in pumping the tank full of oil weighing 50 lb/ft^3 and available at the level of the x-axis.

   Answer: $\dfrac{75\pi}{4}$ ft · lb

10) A spherical tank of radius 100 ft is full of gasoline weighing 40 lb/ft^3. How much work is done in pumping all the gasoline to the level of the top of the tank?

    Answer: $\dfrac{16\pi}{3} \times 10^9$ ft · lb

11) The graph of $y = x^2$, $0 \le x \le 1$, is rotated around the y-axis to form the surface of a tank. Units on the coordinate axes are in feet. How much work is done in pumping the tank full of oil weighing 50 lb/ft^3 if the oil is initially at the level of the x-axis?

    Answer: $\dfrac{70\pi}{3}$ ft · lb

12) A tank is in the shape of the surface generated by rotating the graph of $y = x^4$, $0 \le x \le 1$, around the y-axis. The tank is initially full of oil weighing 60 lb/ft^3. How much work is done in pumping all the oil to the level of the top of the tank? (Units on the coordinate axes are in feet.)

Answer: 16π ft · lb

13) A tank is in the shape of a hemisphere of radius 60 ft and is resting on its flat base (the curved surface is on top). The tank is filled with ethyl alcohol, which weighs 40 lb/ft^3. How much work is done pumping all the alcohol to the level of the top of the tank?

Answer: $216{,}000{,}000\pi$ ft · lb

6.6 Centroids of Plane Regions and Curves
Solve the problem.

1) Find the centroid of the plane region bounded by the curves.
$x = 0, x = 3, y = 0, y = 4$

Answer: $(\frac{3}{2}, 2)$

2) Find the centroid of the plane region bounded by the curves.
$x = 2, x = 4, y = 3, y = 5$

Answer: $(3, 4)$

3) Find the centroid of the plane region bounded by the curves.
$x = -2, x = 4, y = -1, y = 5$

Answer: $(1, 2)$

4) Find the centroid of the plane region bounded by the curves.
$x = 0, y = 0, 2x + y = 2$

Answer: $(\frac{1}{3}, \frac{2}{3})$

5) Find the centroid of the plane region bounded by the curves.
$y = 0, y = 2x, x + y = 3$

Answer: $(\frac{4}{3}, \frac{2}{3})$

6) Find the centroid of the plane region bounded by the curves.
$y = x^2, y = 4$

Answer: $(0, \frac{12}{5})$

7) Find the centroid of the plane region bounded by the curves.
$y = x^2 - 2, y = -4, x = -3, x = 1$

Answer: $(-\frac{21}{13}, -\frac{137}{130}) \approx (-1.6154, -1.054)$

8) Find the centroid of the plane region bounded by the curves.

   $y = x^2$, $y = 8 - x^2$

   Answer: $(0, 4)$

9) Find the centroid of the plane region bounded by the curves.

   $y = 2x$, $y = 15 - x^2$

   Answer: $(-1, \frac{22}{5})$

10) Find the centroid of the plane region (in the first quadrant) bounded by the curves.

    $y = x^3$, $x = y^3$

    Answer: $(\frac{16}{35}, \frac{16}{35})$

11) The region in the first quadrant bounded by the graphs of $y = x$ and $y = \frac{x^2}{2}$ is rotated around the line $y = x$. Find (a) the centroid of the region and (b) the volume of the solid of revolution.

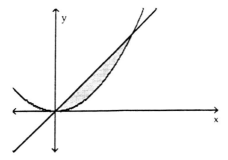

    Answer: (a) $(1, \frac{4}{5})$; (b) $\frac{4}{3}\pi\sqrt{2}$

12) The region in the first quadrant bounded by the graphs of $y = 2x$ and $y = x^2$ is rotated around the line $y = 2x$. Find (a) the centroid of the region and (b) the volume of the solid of revolution.

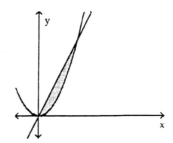

    Answer: (a) $(1, \frac{8}{5})$; (b) $\frac{8}{3}\pi\sqrt{5}$

13) The region in the first quadrant bounded by the graphs of $y = 3x$ and $y = x^3$ is rotated around the line $y = 3x$. Find (a) the centroid of the region and (b) the volume of the solid of revolution.

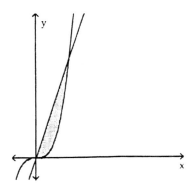

Answer: (a) $(\frac{8\sqrt{3}}{15}, \frac{8\sqrt{3}}{7}) \approx (0.9238, 1.9795)$; (b) $\frac{9}{4}\pi\sqrt{30}$

Ch. 7 Calculus of Transcendental Functions
7.1 Exponential and Logarithmic Functions
Solve the problem.

1) Differentiate the function $f(x) = e^{3x}$.

   Answer: $3e^{3x}$

2) Differentiate the function $f(x) = e^{x^3}$.

   Answer: $3x^2 e^{x^3}$

3) Differentiate the function $f(x) = e^{(1/x^3)}$.

   Answer: $-\dfrac{3e^{(1/x^3)}}{x^4}$

4) Differentiate the function $f(x) = x^3 e^{x^2}$.

   Answer: $3x^2 e^{x^2} + 2x^4 e^{x^2}$

5) Differentiate the function $g(t) = e^{\sin t}$.

   Answer: $e^{\sin t} \cos t$

6) Differentiate the function $f(x) = \dfrac{1 - 2x}{e^x}$.

   Answer: $\dfrac{2xe^x - 3e^x}{e^{2x}}$

7) Differentiate the function $f(x) = \cos(3e^x)$.

   Answer: $-3e^x \sin(3e^x)$

8) Differentiate the function $f(x) =$
   (a) $x\ln x - x$,   (b) $x^5 \ln x$,
   (c) $(\ln x)^2$,   (d) $\dfrac{1-x}{\ln x}$.

   Answer: (a) $\ln|x|$, (b) $x^4 + 5x^4 \ln|x|$, (c) $\dfrac{2\ln|x|}{x}$, (d) $-\dfrac{1}{\ln|x|} - \dfrac{1-x}{x(\ln|x|)^2}$

9) Differentiate the function $f(x) = \ln(2x + 3)$.

   Answer: $\dfrac{2}{2x+3}$

10) Differentiate the function $f(x) = \ln\sqrt{2 + 3x}$.

    Answer: $\dfrac{3}{4 + 6x}$

11) Differentiate the function $f(x) = \ln(\sin x)$.

    Answer: $\cot x$

12) Differentiate the function $f(x) = \sin(\ln x)$.

   Answer: $\dfrac{\cos(\ln x)}{x}$

13) Apply laws of logarithms to simplify the function $f(x) = \ln[(3x-2)^2(x^2-1)^3]$. Then find its derivative.

   Answer: $\dfrac{6}{3x-2} + \dfrac{6x}{x^2-1}$

14) Apply laws of logarithms to simplify the function $f(x) = \ln\sqrt{\dfrac{9-x^2}{4+x^2}}$. Then find its derivative.

   Answer: $\dfrac{13x}{x^4 - 5x^2 - 36}$

15) Apply laws of logarithms to simplify the function $f(x) = \ln\dfrac{x+3}{x-3}$. Then find its derivative.

   Answer: $-\dfrac{6}{x^2-9}$

16) Apply laws of logarithms to simplify the function $g(t) = \ln\dfrac{t^2-1}{t^2}$. Then find its derivative.

   Answer: $\dfrac{2}{t^3 - t}$

17) Given the function $y = 3^x$, find dy/dx by logarithmic differentiation.

   Answer: $3^x \ln 3$

18) Given the function $y = x^{2x}$, find dy/dx by logarithmic differentiation.

   Answer: $x^{2x}(2\ln x + 2)$

19) Given the function $y = \dfrac{(1+2x)^{3/2}}{(1+3x)^{4/3}}$, find dy/dx by logarithmic differentiation.

   Answer: $\dfrac{(1+2x)^{3/2}}{(1+3x)^{4/3}}\left[\dfrac{3}{1+2x} - \dfrac{4}{1+3x}\right]$

20) Given the function $y = (x^4+1)^{x^4}$, find dy/dx by logarithmic differentiation.

   Answer: $(x^4+1)^{x^4}\left[4x^3 \ln(x^4+1) + \dfrac{4x^7}{x^4+1}\right]$

21) Write an equation of the line tangent to the graph of the function $y = 3x^3 \ln x$ at the point $(1, 0)$.

   Answer: $y = 3x - 3$

7.2 Indeterminate Forms and L'Hopital's Rule

Solve the problem.

1) Find $\lim\limits_{x \to \infty} \dfrac{5x - 6}{3x - 4}$. Apply l'Hôpital's rule as many times as necessary, verifying your results after each application.

   Answer: $\dfrac{5}{3}$

2) Find $\lim\limits_{x \to 0} \dfrac{1 - \sec x}{x^3}$. Apply l'Hôpital's rule as many times as necessary, verifying your results after each application.

   Answer: 0

3) Find $\lim\limits_{x \to \infty} \dfrac{e^{2x}}{(x + 5)^3}$. Apply l'Hôpital's rule as many times as necessary, verifying your results after each application.

   Answer: ∞

4) Find $\lim\limits_{x \to \infty} \dfrac{3x}{4x}$. Apply l'Hôpital's rule as many times as necessary, verifying your results after each application.

   Answer: 0

5) Find $\lim\limits_{x \to \infty} \dfrac{\sqrt{4x^3 + 3x}}{\sqrt{2x^3 - 1}}$. Apply l'Hôpital's rule as many times as necessary, verifying your results after each application.

   Answer: $\sqrt{2}$

6) Find $\lim\limits_{x \to \infty} \dfrac{\ln(\ln x)}{x^2 \ln x}$. Apply l'Hôpital's rule as many times as necessary, verifying your results after each application.

   Answer: 0

7) Find $\lim\limits_{x \to 0} \dfrac{12e^x - 6x^2 - 12x - 12}{x^3}$. Apply l'Hôpital's rule as many times as necessary, verifying your results after each application.

   Answer: 2

8) Find $\lim\limits_{x \to 1} \dfrac{x + \cos \pi x}{x^2 - 1}$. Apply l'Hôpital's rule as many times as necessary, verifying your results after each application.

   Answer: $\dfrac{1}{2}$

9) Find $\lim_{x \to 0} \frac{\sqrt{5+2x} - \sqrt{5+x}}{2x}$. Apply l'Hôpital's rule as many times as necessary, verifying your results after each application.

Answer: $\frac{\sqrt{5}}{20}$

10) Find $\lim_{x \to 3} \frac{2x^4 - 3x^3 - 81}{x^5 - 10x^3 + 27}$. Apply l'Hôpital's rule as many times as necessary, verifying your results after each application.

Answer: 1

11) Sketch the graph of $y = \frac{\cos^2 x}{x^2}$, $-2\pi \leq x \leq 2\pi$.

Answer:

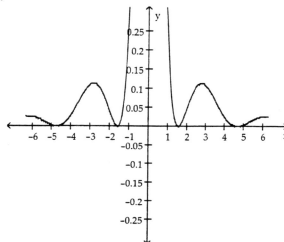

12) Sketch the graph of $y = \frac{\sin x}{x - 2\pi}$, $-2\pi \leq x \leq 2\pi$.

Answer:

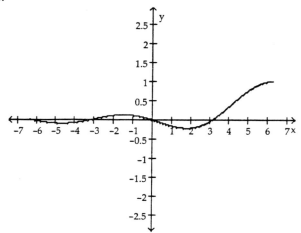

13) Sketch the graph of $y = x^2 e^{-x}$.

Answer:

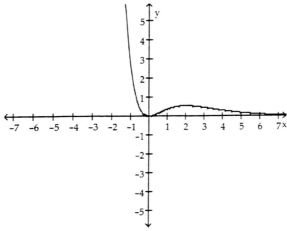

14) Sketch the graph of $y = \dfrac{\ln x}{x^2}$.

Answer:

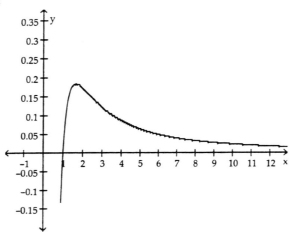

7.3 More Indeterminate Forms
Solve the problem.

1) Find $\lim\limits_{x \to 0} \left(\dfrac{1}{x} - \csc x \right)$

   Answer: 0

2) Find $\lim\limits_{x \to 0} \dfrac{1}{3x} \ln\left(\dfrac{8x + 5}{6x + 5} \right)$

   Answer: $\dfrac{2}{15}$

3) Find $\lim\limits_{x \to \infty} e^{-2x} \ln(2x)$

   Answer: 0

4) Find $\lim_{x \to 4}\left(\dfrac{1}{x-4} - \dfrac{1}{\ln(x-3)}\right)$.

   Answer: $-\dfrac{1}{2}$

5) Find $\lim_{x \to 0^+}\left(\dfrac{1}{\sqrt{2x}} - \cot x\right)$.

   Answer: $-\infty$

6) Find $\lim_{x \to \infty}(\sqrt{x^3+x} - \sqrt{x^3-x})$.

   Answer: 0

7) Find $\lim_{x \to 0} x^{(\tan x)}$.

   Answer: 1

8) Find $\lim_{x \to \infty}\left(\dfrac{3x-2}{3x+2}\right)^x$.

   Answer: $e^{-4/3} = \dfrac{1}{e^{4/3}}$

9) Find $\lim_{x \to \infty}\left(\sin(\dfrac{1}{x^2})\right)^{x^5}$.

   Answer: 0

7.4 The Natural Logarithm as an Integral
Solve the problem.

1) Given $f(x) = 77^{x^3}$, find $f'(x)$.

   Answer: $f'(x) = 3 \cdot 77^{x^3} \cdot x^2 \ln 77$

2) Given $f(x) = 10^{\sin x}$, find $f'(x)$.

   Answer: $f'(x) = 10^{\sin x} \cos x \ln 10$

3) Given $f(x) = 2^{\ln x}$, find $f'(x)$.

   Answer: $f'(x) = \dfrac{2^{\ln x} \ln 2}{x}$

4) Given $f(x) = \log_x 2$, find $f'(x)$.

   Answer: $f'(x) = -\dfrac{\ln 2}{x(\ln x)^2}$

5) Given $f(x) = \log_2(\ln x)$, find $f'(x)$.

   Answer: $f'(x) = \dfrac{1}{x(\ln 2)(\ln x)} = \dfrac{\log_2 e}{x \ln x}$

6) Given $f(x) = \ln 2^x$, find $f'(x)$.

   Answer: $f'(x) = \ln 2$

7) Given $f(x) = x^{\ln x}$, find $f'(x)$.

   Answer: $f'(x) = 2x^{\ln x - 1} \ln x$

8) Given $f(x) = \log_{10} x$, find $f'(x)$.

   Answer: $f'(x) = \dfrac{1}{x \ln 10} = \dfrac{\log_{10} e}{x}$

9) Given $f(x) = (2\pi)^x + x^{(2\pi)} + (2\pi)(2\pi)$, find $f'(x)$.

   Answer: $f'(x) = (2\pi)^x \ln(2\pi) + (2\pi) x^{(2\pi - 1)}$

10) Given $f(x) = \log_4 5^{(2x)}$ find $f'(x)$.

    Answer: $f'(x) = 2 \log_4 5 = 2 \log_4 e \ln 5$

11) Evaluate $\int 6^x \, dx$

    Answer: $\dfrac{6^x}{\ln 6} + C$

12) Evaluate $\displaystyle\int \dfrac{10^{\ln x}}{x} \, dx$

    Answer: $\dfrac{10^{\ln x}}{\ln 10} + C$

13) Evaluate $\int x \cdot 2^{(-x^2)} \, dx$

    Answer: $-\dfrac{2^{(-x^2)}}{2 \ln 2} + C$

14) Evaluate $\displaystyle\int \dfrac{\log_3 x}{2x} \, dx$

    Answer: $\dfrac{(\ln x)^2}{4 \ln 3} + C$

15) Find the lowest point on the curve $f(x) = x 2^x$.

    Answer: $(-1.44269, -0.53074)$

16) Find $\dfrac{dy}{dx}$ if $y = \log_{2x} 4$.

    Answer: $-\dfrac{\ln 4}{x (\ln 2x)^2}$

7.5 Inverse Trigonometric Functions
Solve the problem.

1) Find the values indicated:
   (a) $\sin^{-1}(\frac{\sqrt{3}}{2})$; (b) $\sin^{-1}(1)$; (c) $\sin^{-1}(-\frac{\sqrt{2}}{2})$; (d) $\sin^{-1}(0)$

   Answer: (a) $\frac{\pi}{3}$; (b) $\frac{\pi}{2}$; (c) $-\frac{\pi}{4}$; (d) 0

2) Find the values indicated:
   (a) $\cos^{-1}(\frac{\sqrt{3}}{2})$; (b) $\cos^{-1}(1)$; (c) $\cos^{-1}(-\frac{\sqrt{2}}{2})$; (d) $\cos^{-1}(0)$

   Answer: (a) $\frac{\pi}{6}$; (b) 0; (c) $-\frac{\pi}{4}$; (d) $\frac{\pi}{2}$

3) Find the values indicated:
   (a) $\cot^{-1}(\frac{\sqrt{3}}{3})$; (b) $\cot^{-1}(1)$; (c) $\cot^{-1}(0)$; (d) $\cot^{-1}(-1)$

   Answer: (a) $\frac{\pi}{3}$; (b) $\frac{\pi}{4}$; (c) $\frac{\pi}{2}$; (d) $-\frac{\pi}{4}$

4) Find the values indicated:
   (a) $\csc^{-1}(1)$; (b) $\csc^{-1}(-1)$; (c) $\csc^{-1}(2)$; (d) $\csc^{-1}(\sqrt{2})$

   Answer: (a) $\frac{\pi}{2}$; (b) $-\frac{\pi}{2}$; (c) $\frac{\pi}{6}$; (d) $\frac{\pi}{4}$

5) Given $f(x) = \sin^{-1}(x^3)$, find $f'(x)$.

   Answer: $f'(x) = \dfrac{3x^2}{\sqrt{1-x^6}}$

6) Given $f(x) = \arctan(\sec x)$, find $f'(x)$.

   Answer: $f'(x) = \dfrac{\sec x \tan x}{1 + \sec^2 x}$

7) Given $f(x) = \ln(\tan^{-1} x)$, find $f'(x)$.

   Answer: $f'(x) = \dfrac{1}{(1+x^2)\arctan x}$

8) Given $f(x) = \arcsin(\frac{1}{x})$, find $f'(x)$.

   Answer: $f'(x) = -\dfrac{1}{x\sqrt{x^2-1}}$

9) Given $f(x) = \tan^{-1}(e^{2x})$, find $f'(x)$.

   Answer: $f'(x) = \dfrac{2e^{2x}}{1+e^{4x}}$

10) Given $f(x) = \text{arcsec}(e^x)$, find $f'(x)$.

Answer: $f'(x) = \dfrac{1}{\sqrt{e^{2x}-1}}$

11) Given $f(x) = \sin^{-1}(\ln x)$, find $f'(x)$.

Answer: $f'(x) = \dfrac{1}{x\sqrt{1-(\ln x)^2}}$

12) Given $\cot^{-1}x + \cot^{-1}y = \dfrac{\pi}{4}$, (a) find $\dfrac{dy}{dx}$ by implicit differentiation. Then find (b) the equation of the line tangent to the graph at the point $P(1, 2)$.

Answer: (a) $\dfrac{dy}{dx} = -\dfrac{1+y^2}{1+x^2}$; (b) $5x + 2y = 9$

13) Given $\cos^{-1}x + \cos^{-1}y = \dfrac{\pi}{2}$, (a) find $\dfrac{dy}{dx}$ by implicit differentiation. Then find (b) the equation of the line tangent to the graph at the point $P(0, \dfrac{\sqrt{3}}{2})$.

Answer: (a) $\dfrac{dy}{dx} = -\dfrac{\sqrt{1-y^2}}{\sqrt{1-x^2}}$; (b) $x + 2y = \sqrt{3}$

14) Given $(\sec^{-1}x)(\sec^{-1}y) = \dfrac{\pi}{3}$, (a) find $\dfrac{dy}{dx}$ by implicit differentiation. Then find (b) the equation of the line tangent to the graph at the point $P(\sqrt{2}, \sqrt{2})$.

Answer: (a) $\dfrac{dy}{dx} = -\dfrac{|y|\sqrt{y^2-1}}{|x|\sqrt{x^2-1}} \cdot \dfrac{\sec^{-1}y}{\sec^{-1}x}$; (b) $x + y = 2\sqrt{2}$

15) Given $(\tan^{-1}x)^2 + (\cot^{-1}y)^2 = 1$, (a) find $\dfrac{dy}{dx}$ by implicit differentiation. Then find (b) the equation of the line tangent to the graph at the point $P(1, 0)$.

Answer: (a) $\dfrac{dy}{dx} = \dfrac{1+y^2}{1+x^2} \cdot \dfrac{\tan^{-1}x}{\cot^{-1}y}$; (b) $x - 4y = 1$

16) Evaluate $\displaystyle\int \dfrac{1}{e^{2x}\sqrt{e^{4x}-1}}\,dx$.

Answer: $\dfrac{\sqrt{e^{4x}-1}}{2e^{2x}} + C$

17) Evaluate $\displaystyle\int \dfrac{e^x}{1+e^{2x}}\,dx$.

Answer: $\arctan(e^x) + C$

18) Evaluate $\displaystyle\int \dfrac{1}{4+9x^2}\,dx$.

Answer: $\dfrac{2}{3}\tan^{-1}\left(\dfrac{3x}{2}\right) + C$

19) Evaluate $\int \frac{1}{2+3x^2} dx$.

Answer: $\frac{\sqrt{6}}{3}\tan^{-1}(\frac{x\sqrt{6}}{2}) + C$

20) Evaluate $\int_{3}^{4} \frac{1}{\sqrt{25-x^2}} dx$.

Answer: $5(\arcsin(\frac{4}{5}) - \arcsin(\frac{3}{5}))$

21) Evaluate $\int_{-1}^{1} \frac{1}{1+x^2} dx$.

Answer: $\frac{\pi}{2}$

22) Evaluate $\int \frac{dx}{\sqrt{x(4-x)}}$.

Answer: $2\sin^{-1}(\frac{\sqrt{x}}{2}) + C$

23) Evaluate $\int_{2}^{e} \frac{dx}{x\ln x \sqrt{(\ln x)^2 - 1}}$.

Answer: $-\operatorname{arcsec}(\ln 2)$

24) Evaluate $\int \frac{dx}{4x\sqrt{x^2-16}}$.

Answer: $\frac{1}{4}\sec^{-1}\left|\frac{x}{4}\right| + C$

25) Evaluate $\int \frac{\arcsin(2x)}{\sqrt{1-4x^2}} dx$.

Answer: $\frac{1}{4}(\arcsin(2x))^2 + C$

26) Find the volume generated by revolving around the y-axis the area under $y = \frac{1}{1+x^4}$ from $x = 0$ to $x = 1$

Answer: $\frac{\pi^2}{4}$

7.6 Hyperbolic Functions
Solve the problem.

1) Given $f(x) = \ln(\cosh x)$, find $f'(x)$.

Answer: $f'(x) = \tanh x$

2) Given $f(x) = \cosh(\ln x)$, find $f'(x)$.

　　Answer: $f'(x) = \dfrac{\sinh(\ln x)}{x} = \dfrac{x^2 - 1}{2x^2}$

3) Given $f(x) = e^{\cosh x}$, find $f'(x)$.

　　Answer: $f'(x) = e^{\cosh x} \sinh x$

4) Given $f(x) = \tanh x^2$, find $f'(x)$.

　　Answer: $f'(x) = 2x\operatorname{sech}^2 x^2$

5) Given $f(x) = x^3 \operatorname{sech}(\dfrac{1}{x^2})$, find $f'(x)$.

　　Answer: $f'(x) = 3x^2 \operatorname{sech}(\dfrac{1}{x^2}) + 2\operatorname{sech}(\dfrac{1}{x^2})\tanh(\dfrac{1}{x^2})$

6) Given $f(x) = \dfrac{2}{x^2 + \cosh x^2}$, find $f'(x)$.

　　Answer: $f'(x) = -\dfrac{4x(1 + \sinh x^2)}{(x^2 + \cosh x^2)^2}$

7) Evaluate: $\int \cosh 2x\, dx$

　　Answer: $\dfrac{\sinh 2x}{2} + C$

8) Evaluate: $\int \sinh^6 x \cosh x\, dx$

　　Answer: $\dfrac{\sinh^7 x}{7} + C$

9) Evaluate: $\displaystyle\int \dfrac{\cosh x}{\sqrt{1 - \sinh^2 x}}\, dx$

　　Answer: $\arcsin(\sinh x) + C$

10) Evaluate: $\int x^2 \cosh x^3\, dx$

　　Answer: $\dfrac{\sinh x^3}{3} + C$

11) Evaluate: $\displaystyle\int \dfrac{\sinh\sqrt{x}}{\sqrt{x}}\, dx$

　　Answer: $2\cosh\sqrt{x} + C$

12) Evaluate: $\int 4x^2 \sinh x^3\, dx$

　　Answer: $\dfrac{4}{3}\cosh x^3 + C$

13) Evaluate: $\int \dfrac{\tanh x \,\text{sech}\, x}{1 + \text{sech}\, x}\, dx$

Answer: $-\ln(1 + \text{sech}\, x) + C$

14) Evaluate: $\int x\,\text{csch}^2 x^2 \coth^2 x^2\, dx$

Answer: $-\dfrac{1}{6}\coth^3 x^2 + C$

15) Evaluate: $\int \sinh^2 x\, dx$

Answer: $\dfrac{1}{4}\sinh 2x - \dfrac{1}{2}x + C$

16) Given $f(x) = \sinh^{-1}(e^x)$, find $f'(x)$.

Answer: $f'(x) = \dfrac{e^x}{\sqrt{1 + e^{2x}}}$

17) Given $f(x) = \ln(\sinh^{-1} x)$, find $f'(x)$.

Answer: $f'(x) = \dfrac{1}{\text{arcsinh}\, x \sqrt{1 + x^2}}$

18) Given $f(x) = \cosh(\sinh^{-1} x)$, find $f'(x)$.

Answer: $f'(x) = \dfrac{x}{\sqrt{x^2 + 1}}$

19) Given $f(x) = \tanh^{-1}(\sin x)$, find $f'(x)$.

Answer: $f'(x) = \dfrac{\cos x}{1 - \sin^2 x} = \sec x$

20) Given $f(x) = \text{csch}^{-1}(\dfrac{1}{x^2})$, find $f'(x)$.

Answer: $f'(x) = \dfrac{2x}{\sqrt{x^4 + 1}}$

21) Given $f(x) = \cosh^{-1} e^{2x}$, find $f'(x)$.

Answer: $f'(x) = \dfrac{2e^{2x}}{\sqrt{e^{4x} - 1}}$

22) Given $f(x) = (\tanh^{-1} x^2)^{3/2}$, find $f'(x)$.

Answer: $f'(x) = \dfrac{3x\sqrt{\tanh^{-1} x^2}}{1 - x^4}$

23) Given $f(x) = \ln(\sinh^{-1} 2x)$, find $f'(x)$.

Answer: $f'(x) = \dfrac{2}{(\sinh^{-1} 2x)\sqrt{1 + 4x^2}}$

24) Evaluate: $\int \dfrac{dx}{\sqrt{x^2-4}}$

Answer: $\cosh^{-1}(\dfrac{x}{2}) + C$

25) Evaluate: $\int_{1}^{2} \dfrac{dx}{9-x^2}$

Answer: $\dfrac{1}{6}(\coth^{-1}(\dfrac{2}{3}) - \coth^{-1}(\dfrac{1}{3}))$

26) Evaluate: $\int \dfrac{dx}{x\sqrt{x^2+36}}$

Answer: $-\dfrac{1}{6}\operatorname{csch}^{-1}(\dfrac{x}{6}) + C$

27) Evaluate: $\int \dfrac{e^x}{\sqrt{e^{2x}-1}}\,dx$

Answer: $\cosh^{-1} e^x + C$

28) Evaluate: $\int \dfrac{\sin x}{\sqrt{1+\cos^2 x}}\,dx$

Answer: $-\sinh^{-1}(\cos x) + C$

29) Find the length of the curve $y = \cosh x$ over the interval $[0, \ln 2]$.

Answer: $\dfrac{3}{4}$

30) Find the length of the curve $y = \sinh x$ over the interval $[0, \dfrac{\pi}{a}]$.

Answer: $\cosh(\dfrac{\pi}{a}) - 1$

31) Estimate (graphically or numerically) the points of intersection of the curves $y = x$ and $y = \text{sech}^2 x$. Then approximate the area of the region bounded by these two curves and the y-axis.

Answer: $A \approx 0.3604$

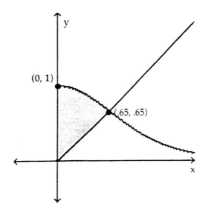

32) Find the volume of the solid obtained by revolving around the x-axis the area under $y = \cosh(\frac{x}{2})$ from $x = 0$ to $x = \pi$.

Answer: $V \approx 23.08$

Ch. 8 Techniques of Integration

8.1 Introduction

1) There are no exercises for this section.

   Answer:

8.2 Integral Tables and Simple Substitutions

Solve the problem.

1) Evaluate $\int x^2\sqrt{5x^3+1}\,dx$.

   Answer: $\dfrac{2}{45}(5x^3+1)^{3/2}+C$

2) Evaluate $\int 6x^3 \csc^2 x^4\,dx$.

   Answer: $-\dfrac{3}{2}\cot(x^4)+C$

3) Evaluate $\displaystyle\int \dfrac{t^2}{\sqrt{16t^3+5}}\,dt$.

   Answer: $\dfrac{\sqrt{16t^3+5}}{24}+C$

4) Evaluate $\displaystyle\int \dfrac{dx}{\sqrt{4-x^2}}$.

   Answer: $\sin^{-1}\left(\dfrac{x}{2}\right)+C$

5) Evaluate $\displaystyle\int \dfrac{1}{1+9t^2}\,dt$.

   Answer: $\dfrac{\tan^{-1}(3t)}{3}+C$

6) Evaluate $\displaystyle\int \dfrac{(\sqrt{x}+4)^3}{3\sqrt{x}}\,dx$.

   Answer: $\dfrac{(\sqrt{x}+4)^4}{6}+C$

7) Evaluate $\displaystyle\int \dfrac{\csc^2\theta}{1+\cot\theta}\,d\theta$.

   Answer: $-\ln|1+\cot\theta|+C$

8) Evaluate $\displaystyle\int \dfrac{2x}{\sqrt{x-1}}\,dx$.

   Answer: $\dfrac{4}{3}(x+2)\sqrt{x-1}+C$

9) Evaluate $\int 5x\sqrt{x+1}\,dx$.

   Answer: $\dfrac{2}{3}(x+1)^{3/2}(3x-2)+C$

10) Use the fact that $4x^2 + 4x + 2 = (2x+1)^2 + 1$ to evaluate $\int \frac{1}{4x^2 + 4x + 2} dx$.

Answer: $\frac{1}{2}\tan^{-1}(2x+1) + C$

11) Use the fact that $4 - (x-2)^2 = 4x - x^2$ to evaluate $\int \frac{1}{\sqrt{4x - x^2}} dx$.

Answer: $\sin^{-1}(\frac{x-2}{2}) + C$

8.3 Integration by Parts
Solve the problem.

1) Evaluate $\int 2xe^x \, dx$.

   Answer: $2e^x(x-1) + C$

2) Evaluate $\int t^3 \cos t \, dt$.

   Answer: $(t^3 - 6t)\sin t + 3(t^2 - 2)\cos t + C$

3) Evaluate $\int \arcsin x \, dx$.

   Answer: $x \arcsin x + \sqrt{1-x^2} + C$

4) Evaluate $\int \frac{\ln(2x)}{x^2} dx$.

   Answer: $\frac{-(1 + \ln 2 + \ln x)}{x} + C$

5) Evaluate $\int 3x\sqrt{4x+3} \, dx$.

   Answer: $\frac{3(2x-1)(4x+3)^{3/2}}{20} + C$

6) Evaluate $\int x^2 \sqrt{1-x^2} \, dx$.

   Answer: $\frac{\sin^{-1} x + x(2x^2 - 1)\sqrt{1-x^2}}{8} + C$

7) Evaluate $\int x^3 \arctan x \, dx$.

   Answer: $\frac{(x^4 - 1)\arctan x}{4} - \frac{x(x^2 - 3)}{12} + C$

8) Evaluate $\int x^2 \sin 2x \, dx$.

   Answer: $\frac{(1 - 2x^2)\cos 2x}{4} + \frac{x \sin 2x}{2} + C$

9) Evaluate $\int 2x \tan^{-1}(4x) \, dx$.

   Answer: $\frac{(16x^2 + 1)\tan^{-1}(4x) - 4x}{16} + C$

Calculus 175

10) Evaluate $\int x\sinh x\, dx$.

   Answer: $x\cosh x - \sinh x + C$

11) Evaluate $\int x^3 \cos x^2\, dx$ by first making a substitution of the form $t = x^k$ and then integrate by parts.

   Answer: $\dfrac{\cos x^2 + x^2 \sin x^2}{2} + C$

12) Evaluate $\int x^2 \cos x^{3/2}\, dx$ by first making a substitution of the form $t = x^k$ and then integrate by parts.

   Answer: $\dfrac{2(\cos x^{3/2} + x^{3/2}\sin x^{3/2})}{3} + C$

13) R is bounded below by the x-axis and above by the curve $y = 2\cos x$, $0 \le x \le \dfrac{\pi}{2}$. Find the volume of the solid generated by revolving R around the y-axis by the method of cylindrical shells.

   Answer: $2\pi^2 - 4\pi$

14) R is bounded on the left by the y-axis, on the right by the line $x = 1$, above by the curve $y = e^x$ and below by the curve $y = x^2$. Find the volume of the solid generated by revolving R around the y-axis by the method of cylindrical shells.

   Answer: $\dfrac{3\pi}{2}$

15) Use integration by parts to evaluate $\int xe^x \sin x\, dx$.

   Answer: $\dfrac{-e^x((x-1)\cos x - x\sin x)}{2} + C$

16) Use integration by parts to evaluate $\int \sin 2x \cos 3x\, dx$.

   Answer: $\dfrac{5\cos x - \cos 5x}{10} + C$

17) Find the area and the centroid of the region R that is bounded by the curves $y = x^4$ and $y = 4^x$ for $2 \le x \le 4$.

   Answer: $A \approx 25.2766$; Centroid $\approx (3.21663, 109.3)$

8.4 Trigonometric Integrals
Solve the problem.

1) Evaluate $\int \sin^3 x\, dx$.

   Answer: $\dfrac{\cos^3 x}{3} - \cos x + C$

2) Evaluate $\int \sin^2 x \cos^2 x\, dx$.

   Answer: $\dfrac{x}{8} - \dfrac{\sin 4x}{32} + C$

3) Evaluate $\int x\sec^2 x\, dx$.

   Answer: $x\tan x - \ln|\sec x| + C$

4) Evaluate $\int \tan^3 x \, dx$.

   Answer: $\dfrac{\tan^2 x}{2} + \ln|\cos x| + C$

5) Evaluate $\int \cos(\ln x) \, dx$.

   Answer: $\dfrac{x}{2}[\cos(\ln|x|) + \sin(\ln|x|)] + C$

6) Evaluate $\int \tan^3 u \sec^3 u \, du$.

   Answer: $\dfrac{\sec^5 u}{5} - \dfrac{\sec^3 u}{3} + C$

7) Evaluate $\int \tan^2 x \sec^4 x \, dx$.

   Answer: $\dfrac{\tan^3 x}{3} + \dfrac{\tan^5 x}{5} + C$

8) Evaluate $\displaystyle\int \dfrac{\cos x}{1 + \sin^2 x} \, dx$.

   Answer: $\arctan(\sin x) + C$

9) Evaluate $\int \sin^2 3x \, dx$.

   Answer: $\dfrac{x}{2} - \dfrac{\sin 6x}{12} + C$

10) Evaluate $\int \sec^4 x \tan^3 x \, dx$.

    Answer: $\dfrac{\sec^6 x}{6} - \dfrac{\sec^4 x}{4} + C$

11) Evaluate $\int e^{-x} \cos x \, dx$.

    Answer: $\dfrac{1}{2e^x}(\sin x - \cos x) + C$

12) Evaluate $\int \sin^2 \omega \, d\omega$.

    Answer: $\dfrac{1}{2}(\omega - \sin\omega \cos\omega) + C = \dfrac{\omega}{2} - \dfrac{\sin 2\omega}{4} + C$

13) Evaluate $\displaystyle\int \dfrac{\sec^4 x}{\tan^4 x} \, dx$.

    Answer: $-\cot x - \dfrac{\cot^3 x}{3} + C$

14) Evaluate $\int \sin^5 x \, dx$.

    Answer: $-\cos x + \dfrac{2\cos^3 x}{3} - \dfrac{\cos^5 x}{5} + C$

15) Evaluate $\int \tan^2 x \sec^4 x \, dx$.

Answer: $\dfrac{\sec^4 x \tan x}{5} + C$

16) Evaluate $\int \sin^4 x \, dx$.

Answer: $\dfrac{3x}{8} - \dfrac{\sin 2x}{4} + \dfrac{\sin 4x}{32} + C$

17) Evaluate $\int \tan^3 x \sec^3 x \, dx$.

Answer: $\dfrac{\sec^5 x}{5} - \dfrac{\sec^3 x}{3} + C$

18) Evaluate $\displaystyle\int \dfrac{1}{1 + 3\sin z + \cos z} \, dz$.

Answer: $\dfrac{1}{3}\ln|1 + 3\tan z| + C$

19) Evaluate $\int \sin^2 x \cos^3 x \, dx$.

Answer: $\dfrac{\sin 2x}{4} - \dfrac{\sin 8x}{16} + C$

20) Evaluate $\displaystyle\int \dfrac{1}{\sec x \tan x} \, dx$.

Answer: $\cos x + \ln|\csc x - \cot x| + C$

21) Evaluate $\int x \cos 3x \, dx$.

Answer: $\dfrac{\cos 3x}{9} + \dfrac{x \sin 3x}{3} + C$

22) Evaluate $\int \sin\sqrt{x} \, dx$.

Answer: $2\sin\sqrt{x} - 2\sqrt{x}\cos\sqrt{x} + C$

23) Find the area of the region R bounded by the x-axis and the curve $y = 2\sin^2 x$, from $x = 0$ to $x = \pi$.

Answer: $A = \displaystyle\int_0^\pi 2\sin^2 x \, dx = \pi$

24) Find the area of the region R bounded by the curves $y = \cos^2 x$ and $y = \sin x \cos x$, from $x = \dfrac{\pi}{4}$ to $x = \dfrac{5\pi}{4}$.

Answer: $A = \displaystyle\int_{\pi/4}^{5\pi/4} (\cos^2 x - \sin x \cos x) \, dx = \dfrac{\pi}{2}$

25) Find the volume of the solid formed by rotating the region R, bounded by the x-axis, the curve $y = \cos^2 x + 1$ and the lines $x = -\pi$ and $x = \pi$, around the x-axis.

Answer: $\dfrac{19\pi^2}{4}$

26) Find the volume of the solid formed by rotating the region R, bounded by the x-axis, the y-axis, the curve $y = \tan x$ and the lines $x = 0$ and $x = \frac{\pi}{4}$, around the x-axis.

Answer: $\dfrac{4\pi - \pi^2}{4}$

27) Find the length of the graph of $y = \ln(\sin x)$ from $x = \dfrac{\pi}{2}$ to $x = \dfrac{2\pi}{3}$.

Answer: $\dfrac{\ln 3}{2}$

28) Find $\int \sin 4x \cos x \, dx$ using the trigonometric identities:
$\sin A \sin B = \dfrac{1}{2}[\cos(A - B) - \cos(A + B)]$,
$\sin A \cos B = \dfrac{1}{2}[\sin(A - B) + \sin(A + B)]$,
$\cos A \cos B = \dfrac{1}{2}[\cos(A - B) + \cos(A + B)]$.

Answer: $\dfrac{-3\cos 3x}{6} - \dfrac{\cos 5x}{10} + C$

29) Find $\int \cos 2x \cos 4x \, dx$ using the trigonometric identities:
$\sin A \sin B = \dfrac{1}{2}[\cos(A - B) - \cos(A + B)]$,
$\sin A \cos B = \dfrac{1}{2}[\sin(A - B) + \sin(A + B)]$,
$\cos A \cos B = \dfrac{1}{2}[\cos(A - B) + \cos(A + B)]$.

Answer: $\dfrac{\sin 6x}{12} + \dfrac{\sin 2x}{4} + C$

30) Find $\int \sin 3x \sin 4x \, dx$ using the trigonometric identities:
$\sin A \sin B = \dfrac{1}{2}[\cos(A - B) - \cos(A + B)]$,
$\sin A \cos B = \dfrac{1}{2}[\sin(A - B) + \sin(A + B)]$,
$\cos A \cos B = \dfrac{1}{2}[\cos(A - B) + \cos(A + B)]$.

Answer: $\dfrac{\sin x}{2} - \dfrac{\sin 7x}{14} + C$

8.5 Rational Functions and Partial Fractions
Solve the problem.

1) Evaluate $\int \dfrac{4}{x^2 - 1} \, dx$.

Answer: $-2\ln\left|\dfrac{x + 1}{x - 1}\right| + C$

2) Evaluate $\int \dfrac{2x^3 - 12x - 15}{x^4 + 4x^3 + 5x^2}\, dx$.

Answer: $\ln|x^2 + 4x + 5| - \arctan(2 + x) + \dfrac{3}{x} + C$

3) Evaluate $\int \dfrac{5x^3 + x^2 - 3x - 3}{x^3(x + 1)}\, dx$.

Answer: $\dfrac{3}{2x^2} + \ln|x| + 4\ln|x + 1| + C$

4) Evaluate $\int \dfrac{3x + 5}{x^4 + x^2}\, dx$.

Answer: $\dfrac{3}{2}\ln\left|\dfrac{x^2}{x^2 + 1}\right| - 5\arctan x - \dfrac{5}{x} + C$

5) Evaluate $\int \dfrac{2x^2}{(x^2 + 4)(x - 2)}\, dx$.

Answer: $\dfrac{1}{2}\ln|(x - 2)^2(x^2 + 4)| + \arctan\left(\dfrac{x}{2}\right) + C$

6) Evaluate $\int \dfrac{x^5 + 1}{x^2}\, dx$.

Answer: $\dfrac{x^4}{4} - \dfrac{1}{x} + C$

7) Evaluate $\int \dfrac{1}{x^2 + 2x}\, dx$.

Answer: $\dfrac{1}{2}\ln\left|\dfrac{x}{x + 2}\right| + C$

8) Evaluate $\int \dfrac{3x + 3}{x^3 - 1}\, dx$.

Answer: $2\ln(x - 1) - \ln|x^2 + x + 1| + C$

9) Evaluate $\int \dfrac{3x^3 + 3x^2 + x - 2}{x^4 + x^2}\, dx$.

Answer: $\ln(x^2 + 1) + \ln|x| + 5\arctan x + C$

10) Evaluate $\int \dfrac{x}{(x + 2)^2}\, dx$.

Answer: $\ln|x + 2| - \dfrac{x}{x + 2} + C$

11) Evaluate $\int \dfrac{x^2 + 5x - 11}{x^3 + x^2 + 4x + 4}\, dx$.

Answer: $\dfrac{1}{2}\tan^{-1}(\dfrac{x}{2}) - 3\ln|x + 1| + 2\ln|x^2 + 4| + C$

12) Evaluate $\int \dfrac{x^5}{1 + x^2}\, dx$.

Answer: $\dfrac{1}{2}\ln(x^2 + 1) + \dfrac{x^2}{4}(x^2 - 2) + C$

13) Evaluate $\int \dfrac{x^3 + 4x^2 + 10x + 3}{x^2 + 3x + 2}\, dx$.

Answer: $9\ln|x + 2| - 4\ln(x + 1) + \dfrac{x^2}{2} + x + C$

14) Evaluate $\int \dfrac{1}{3 - 2x - x^2}\, dx$.

Answer: $\dfrac{1}{4}\ln\left|\dfrac{x + 3}{x - 1}\right| + C$

15) Evaluate $\int \dfrac{1}{x^4 + x^2}\, dx$.

Answer: $-\arctan x - \dfrac{1}{x} + C$

16) Evaluate $\int \dfrac{2e^{4t}}{(2e^{2t} - 2)^3}\, dt$ by making a preliminary substitution and then using the method of partial fractions.

Answer: $\dfrac{1 - 2e^{2t}}{16(e^{2t} - 1)^2} + C$

17) Evaluate $\int \dfrac{\sin\theta}{\cos^2\theta - \cos\theta - 12}\, d\theta$ by making a preliminary substitution and then using the method of partial fractions.

Answer: $\dfrac{1}{7}(\ln|\cos\theta + 3| - \ln|\cos\theta - 4|) + C$

18) Evaluate $\int \dfrac{\ln t + 1}{t(4 + 3\ln t)^2}\, dt$ by making a preliminary substitution and then using the method of partial fractions.

Answer: $\dfrac{1}{9}(\ln(3\ln t + 4) + \dfrac{1}{9(3\ln t + 4)} + C$

19) Evaluate $\int \dfrac{\cos^2 t}{\sin^3 t + \sin^2 t}\, dt$ by making a preliminary substitution and then using the method of partial fractions.

Answer: $\ln|\cos t + 1| - \ln|\sin t| - \cot t + C$

20) R is the region between the curve $y = \dfrac{x-25}{x^2-5x}$ and the x-axis on $1 \le x \le 4$. Find (a) the area of R, (b) the volume obtained by rotating R around the y-axis, and (c) the volume obtained by rotating R around the x-axis.

Answer: (a) $18\ln 2$; (b) $2\pi(40\ln 2 + 3)$; (c) $\dfrac{\pi}{4}(128\ln 2 + 123)$

21) R is the region between the curve $y = \dfrac{x+6}{12+4x-x^2}$ and the x-axis on $0 \le x \le 2$. Find (a) the area of R, (b) the volume obtained by rotating R around the y-axis, and (c) the volume obtained by rotating R around the x-axis.

Answer: (a) $\dfrac{1}{2}\ln(\dfrac{27}{4})$; (b) $9\pi(9\ln 3 - 2(5\ln 2 + 1))$; (c) $\dfrac{\pi}{16}(3\ln 3 + 4)$

22) The plane region R shown below is bounded by the curve $y^2 = \dfrac{4-x}{4+x^2}x^2$ $0 \le x \le 4$. Find volume generated by rotating R around the x-axis.

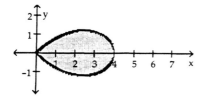

Answer: $\dfrac{128\pi}{3}(3\ln 2 - 2)$

8.6 Trigonometric Substitution
Solve the problem.

1) Use trigonometric substitution to evaluate $\displaystyle\int \dfrac{x}{\sqrt{1-x^2}}\,dx$.

Answer: $-\sqrt{1-x^2} + C$

2) Use trigonometric substitution to evaluate $\displaystyle\int \dfrac{4x^2}{x^4-1}\,dx$.

Answer: $2\arctan x + \ln|x-1| - \ln|x+1| + C$

3) Use trigonometric substitution to evaluate $\displaystyle\int \dfrac{\sqrt{x^2+1}}{x^4}\,dx$.

Answer: $-\sqrt{x^2+1}\left(\dfrac{1}{3x^3} + \dfrac{1}{3x}\right) + C$

4) Use trigonometric substitution to evaluate $\displaystyle\int \dfrac{x^3}{\sqrt{x^2+1}}\,dx$.

Answer: $\dfrac{(x^2-2)\sqrt{x^2+1}}{3} + C$

5) Use trigonometric substitution to evaluate $\int \dfrac{x^3}{(x^2+4)^{3/2}}\,dx$.

Answer: $\dfrac{(x^2+8)}{\sqrt{x^2+4}} + C = \dfrac{(x^2+8)\sqrt{x^2+4}}{x^2+4} + C$

6) Use trigonometric substitution to evaluate $\int \dfrac{x}{\sqrt{x^2+1}}\,dx$.

Answer: $\sqrt{x^2+1} + C$

7) Use trigonometric substitution to evaluate $\int \sqrt{9-x^2}\,dx$.

Answer: $\dfrac{9}{2}\arcsin\left(\dfrac{x}{3}\right) + \dfrac{x}{2}\sqrt{9-x^2} + C$

8) Use trigonometric substitution to evaluate $\int \dfrac{2x^3}{\sqrt{x^4+1}}\,dx$.

Answer: $\sqrt{x^4+1} + C$

9) Use trigonometric substitution to evaluate $\int \dfrac{1}{x\sqrt{x^2-1}}\,dx$.

Answer: $\ln(\sqrt{x^2-1}) + x + C$ or $\arctan(\sqrt{x^2-1}) + C$

10) Use trigonometric substitution to evaluate $\int \dfrac{x}{\sqrt{x^2-16}}\,dx$.

Answer: $\sqrt{x^2-16} + C$

11) Use trigonometric substitution to evaluate $\int \dfrac{1}{\sqrt{1+x^2}}\,dx$.

Answer: $\ln\left|\sqrt{x^2+1} + x\right| + C$

12) Use trigonometric substitution to evaluate $\int \dfrac{x}{\sqrt{x^2+9}}\,dx$.

Answer: $\sqrt{x^2+9} + C$

13) Use trigonometric substitution to evaluate $\int \dfrac{1}{\sqrt{2x-x^2}}\,dx$.

Answer: $\arcsin(x-1) + C$

14) Use trigonometric substitution to evaluate $\int \dfrac{x^3}{x^8+1}\,dx$.

Answer: $\dfrac{1}{4}\arctan(x^4) + C$

15) Use trigonometric substitution to evaluate $\int \frac{1}{(x^2+1)^2} dx$.

Answer: $\frac{1}{2}\tan^{-1}x + \frac{x}{2(x^2+1)} + C$

16) Use trigonometric substitution to evaluate $\int \sqrt{2-x^2}\, dx$.

Answer: $\sin^{-1}(\frac{x\sqrt{2}}{2}) + \frac{x}{2}\sqrt{2-x^2} + C$

17) Use trigonometric substitution to evaluate $\int x^3\sqrt{x^2+4}\, dx$.

Answer: $\frac{(x^2+4)^{3/2}(3x^2-8)}{15} + C$

18) Use trigonometric substitution to evaluate $\int \frac{\sqrt{x^2+1}}{x}\, dx$.

Answer: $\ln\left(\frac{|\sqrt{x^2+1}-1|}{|x|}\right) + \sqrt{x^2+1} + C$

19) Use hyperbolic substitution to evaluate $\int \frac{1}{\sqrt{9+x^2}}\, dx$.

Answer: $\sinh^{-1}(\frac{x}{3}) + C$

20) Use hyperbolic substitution to evaluate $\int \sqrt{4+x^2}\, dx$.

Answer: $x\sqrt{4+x^2} + 2\sinh^{-1}(\frac{x}{2}) + C$

21) Use hyperbolic substitution to evaluate $\int \frac{\sqrt{x^2-25}}{x^2}\, dx$.

Answer: $\cosh^{-1}(\frac{x}{5}) - \frac{\sqrt{x^2-25}}{x} + C$

22) Use hyperbolic substitution to evaluate $\int \frac{1}{\sqrt{1+16x^2}}\, dx$.

Answer: $\frac{1}{4}\sinh^{-1}(4x) + C$

23) Use hyperbolic substitution to evaluate $\int 4x^2\sqrt{1+4x^2}\, dx$.

Answer: $\frac{x}{8}(8x^2+1)\sqrt{4x^2+1} - \frac{1}{16}\sinh^{-1}(2x) + C$

24) Given the curve $f(x) = 2x^2$ on the interval $[0, 2]$, compute (a) the arc length and (b) the area of the surface obtained by rotating f around the x-axis.

Answer: (a) ≈ 8.40932; (b) ≈ 204.142

25) Given the curve $f(x) = \ln(2x)$ on the interval $[1, 4]$, compute (a) the arc length and (b) the area of the surface obtained by rotating f around the y-axis.

Answer: (a) ≈ 3.3428; (b) ≈ 21.0034

8.7 Integrals Involving Quadratic Polynomials
Solve the problem.

1) Evaluate $\int \dfrac{1}{\sqrt{3 - 2x - x^2}} \, dx$.

Answer: $\arcsin(\dfrac{x + 1}{2}) + C$

2) Evaluate $\int \dfrac{1}{\sqrt{x^2 + 4x + 2}} \, dx$.

Answer: $\dfrac{1}{2}\arcsin(2x + 1) + C$

3) Evaluate $\int \dfrac{1}{\sqrt{3 + 2x - x^2}} \, dx$.

Answer: $\arcsin(\dfrac{x - 1}{2}) + C$

4) Evaluate $\int \dfrac{1}{x^2 + 4x + 13} \, dx$.

Answer: $\dfrac{1}{3}\arctan(\dfrac{x + 2}{3}) + C$

5) Evaluate $\int \dfrac{1}{x^2 + 4x + 5} \, dx$.

Answer: $\tan^{-1}(x + 2) + C$

6) Evaluate $\int \dfrac{x^2 + 5x - 11}{x^3 + x^2 + 4x + 4} \, dx$.

Answer: $\dfrac{1}{2}\tan^{-1}(\dfrac{x}{2}) + 2\ln(x^2 + 4) - 3\ln|x + 1| + C$

7) Evaluate $\int \dfrac{1}{\sqrt{3 + 4x - 4x^2}} \, dx$.

Answer: $\dfrac{1}{2}\arcsin(\dfrac{2x - 1}{2}) + C$

8) Evaluate $\int \dfrac{x^2 + 3x + 1}{x^3 + x} \, dx$.

Answer: $3\tan^{-1}x + \ln|x| + C$

9) Evaluate $\int \sqrt{5 - 4x - x^2} \, dx$.

Answer: $\dfrac{x + 2}{2}\sqrt{5 - 4x - x^2} + \dfrac{9}{2}\arcsin(\dfrac{2 + x}{3}) + C$

10) Evaluate $\int \sqrt{x^2 - 2x}\, dx$.

Answer: $\dfrac{x-1}{2}\sqrt{x^2 - 2x} - \dfrac{1}{2}\ln(\sqrt{x^2 - 2x} + x - 1) + C$

11) R is the region that lies between the curve $y = \dfrac{1}{x^2 + 4x + 5}$ and the x-axis from $x = -3$ to $x = -1$. Find (a) the area of R, (b) the volume of the solid generated by revolving R around the y-axis, and (c) the volume of the solid generated by revolving R around the x-axis.

Answer: (a) $\dfrac{\pi}{2}$; (b) $-2\pi^2$; (c) $\dfrac{\pi}{4}(\pi + 2)$

12) R is the region that lies between the curve $y = \dfrac{1}{9x^2 - 18x + 25}$ and the x-axis from $x = -1$ to $x = 3$. Find (a) the area of R, (b) the volume of the solid generated by revolving R around the y-axis, and (c) the volume of the solid generated by revolving R around the x-axis.

Answer: (a) ≈ 0.163799; (b) ≈ 1.02918; (c) ≈ 0.023633

13) Given $\int \dfrac{x - 2}{x^3 - x^2 - 4}\, dx$, factor the denominator by noting a root r and then employing long division by $x - r$. Then use the method of partial fractions to aid in finding the antiderivative.

Answer: $\dfrac{2\sqrt{7}}{7}\tan^{-1}\left[\dfrac{\sqrt{7}}{7}(2x + 1)\right] + C$

14) Given $\int \dfrac{2x^2 + 3x - 2}{x^3 + 8}\, dx$, factor the denominator by noting a root r and then employing long division by $x - r$. Then use the method of partial fractions to aid in finding the antiderivative.

Answer: $\ln|x^2 - 2x + 4| + \dfrac{\sqrt{3}}{3}\tan^{-1}\left[\dfrac{\sqrt{3}}{3}(x - 1)\right] + C$

15) Given $\int \dfrac{x^4 + x^3}{x^3 + 1}\, dx$, factor the denominator by noting a root r and then employing long division by $x - r$. Then use the method of partial fractions to aid in finding the antiderivative.

Answer: $x + \dfrac{x^2}{2} - \dfrac{2\sqrt{3}}{3}\tan^{-1}\left[\dfrac{\sqrt{3}}{3}(2x - 1)\right] + C$

8.8 Improper Integrals
Solve the problem.

1) Determine whether $\displaystyle\int_{-1}^{\infty} e^{-x}\, dx$ converges or diverges. If it does converge, evaluate the integral.

Answer: converges to e

2) Determine whether $\displaystyle\int_{1}^{\infty} \dfrac{1}{2x\sqrt{x}}\, dx$ converges or diverges. If it does converge, evaluate the integral.

Answer: converges to 1

3) Determine whether $\int_0^8 \dfrac{1}{2x\sqrt{x}}\, dx$ converges or diverges. If it does converge, evaluate the integral.

Answer: diverges to ∞

4) Determine whether $\int_2^\infty \dfrac{1}{x \ln x}\, dx$ converges or diverges. If it converges, evaluate the integral.

Answer: diverges to ∞

5) Determine whether $\int_3^\infty \dfrac{1}{(x-2)^{3/2}}\, dx$ converges or diverges. If it does converge, evaluate the integral.

Answer: converges to 2

6) Determine whether $\int_0^2 \dfrac{x}{\sqrt{4-x^2}}\, dx$ converges or diverges. If it converges, evaluate the integral.

Answer: converges to 2

7) Determine whether $\int_0^{16} \dfrac{1}{(16-x)^{3/2}}\, dx$ converges or diverges. If it does converge, evaluate the integral.

Answer: diverges to ∞

8) Determine whether $\int_{-\infty}^{\infty} \dfrac{1}{1+x^2}\, dx$ converges or diverges. If it converges, evaluate the integral.

Answer: converges to π

9) Determine whether $\int_{-\infty}^{\infty} \dfrac{x}{25+x^2}\, dx$ converges or diverges. If it converges, evaluate the integral.

Answer: does not exist

10) Determine whether $\int_0^{16} \dfrac{1}{\sqrt[4]{x}}\, dx$ converges or diverges. If it does converge, evaluate the integral.

Answer: converges to $\dfrac{32}{3}$

11)

Determine whether $\displaystyle\int_{1/\pi}^{\infty} \frac{1}{x^2} \sin\frac{1}{x}\, dx$ converges or diverges. If it converges, evaluate the integral.

Answer: converges to 2

12)

Determine whether $\displaystyle\int_{3}^{5} \frac{1}{\sqrt{x^2-9}}\, dx$ converges or diverges. If it converges, evaluate the integral.

Answer: converges to ln 3

13)

Determine whether $\displaystyle\int_{0}^{\infty} \frac{x}{1+x^4}\, dx$ converges or diverges. If it converges, evaluate the integral.

Answer: converges to $\dfrac{\pi}{4}$

14)

Determine whether $\displaystyle\int_{0}^{\infty} te^{-t}\, dt$ converges or diverges. If it converges, evaluate the integral.

Answer: converges to 1

15)

Determine whether $\displaystyle\int_{0}^{\infty} \frac{e^{\sqrt[3]{x}}}{\sqrt[3]{x}}\, dx$ converges or diverges. If it converges, evaluate the integral.

Answer: converges to 3

16)

Determine whether $\displaystyle\int_{0}^{1} \frac{1}{x^{2/3}}\, dx$ converges or diverges. If it converges, evaluate the integral.

Answer: converges to 3

17)

Determine whether $\displaystyle\int_{0}^{\infty} 3xe^{-4x}\, dx$ converges or diverges. If it does converge, evaluate the integral.

Answer: converges to $\dfrac{3}{16}$

18)

Determine whether $\int_0^\infty \dfrac{x+1}{x^2+x-1}\, dx$ converges or diverges. If it converges, evaluate the integral.

Answer: diverges to ∞

19)

Determine whether $\int_0^2 \dfrac{1}{(x-1)^{4/5}}\, dx$ converges or diverges. If it converges, evaluate the integral.

Answer: converges to 10

20)

Determine whether $\int_0^\infty \sin x\, dx$ converges or diverges. If it converges, evaluate the integral.

Answer: $\int_0^\infty \sin x\, dx = \lim\limits_{N \to \infty} \int_0^N \sin x\, dx = \lim\limits_{N \to \infty} (1 - \cos N)$

since $\lim\limits_{N \to \infty} \cos N$ does not exist, the integral diverges.

21)

Determine whether $\int_0^{\pi/2} \dfrac{\sin x}{\sqrt{\cos x}}\, dx$ converges or diverges. If it converges, evaluate the integral.

Answer: converges to 2

22)

Investigate the convergence of $\int_0^\infty \dfrac{1}{x^3 + x^2}\, dx$ by expressing it as a sum of two integrals [0, 1] and [1, ∞].

Evaluate the parts that converge.

Answer: $\int_0^1 \dfrac{1}{x^3 + x^2}\, dx = \infty$; $\int_1^\infty \dfrac{1}{x^3 + x^2}\, dx$ converges to $1 - \ln 2$

23)

Investigate the convergence of $\int_0^\infty \dfrac{1}{x^{3/2} + x^{5/2}}\, dx$ by expressing it as a sum of two integrals [0, 1] and [1, ∞].

Evaluate the parts that converge.

Answer: $\int_0^1 \dfrac{1}{x^{3/2} + x^{5/2}}\, dx = \infty$; $\int_1^\infty \dfrac{1}{x^{3/2} + x^{5/2}}\, dx$ converges to $2 - \dfrac{\pi}{2}$

24) Find the area in the first quadrant bounded above by the graph of $f(x) = e^{-x}$ and below by the x-axis.

Answer: 1

Calculus 189

25) Find the area between the graph of $f(x) = xe^{-x^2}$ and the x-axis for $x \geq 0$.

Answer: $\dfrac{1}{2}$

26) Find the area between the x-axis and the graph of $y = (x + 1)^{-3}$ for $x \geq 2$.

Answer: $\dfrac{1}{18}$

27) Let R denote the region bounded below by the x-axis and above by the graph of $y = \dfrac{1}{x^2 + 1}$. Find the volume generated by rotating R around the x-axis.

Answer: $\dfrac{\pi^2}{2}$

Ch. 9 Differential Equations
9.1 Simple Equations and Models
Solve the problem.

1) First find a general solution of the differential equation $\frac{dy}{dx} = 3y$. Then find a particular solution that satisfies the initial condition that $y(1) = 4$.

   Answer: $y(x) = Ae^{3x}$; $y(x) = 4e^{3x} - 3$

2) First find a general solution of the differential equation $\frac{dy}{dx} = 3y^2$. Then find a particular solution that satisfies the initial condition $y(3) = -\frac{1}{9}$.

   Answer: $y(x) = -\frac{1}{3(x+C)}$; $y(x) = -\frac{1}{3x}$

3) First find a general solution of the differential equation $\frac{dy}{dx} = \frac{3}{y}$. Then find a particular solution that satisfies the initial condition $y(0) = 5$.

   Answer: $y(x) = \sqrt{6x + C}$; $y(x) = \sqrt{6x + 25}$

4) First find a general solution of the differential equation $\frac{dy}{dx} = 5y^{3/4}$. Then find a particular solution that satisfies the initial condition that $y(1) = \frac{6561}{256}$

   Answer: $y(x) = (\frac{5}{4}x + C)^4$; $y(x) = (\frac{5}{4}x + 1)^4$

5) Write a differential equation of the form $\frac{dy}{dx} = F(x, y)$ having g as a solution, if the slope of the graph of g at the point (x, y) is the quotient of x and y.

   Answer: $\frac{dy}{dx} = \frac{x}{\sqrt{x^2 + C}}$

6) Write a differential equation that is a mathematical model of the situation in which the rate of change of an organism O = O(t) is proportional to the cube root of O.

   Answer: $\frac{dO}{dt} = k\sqrt[3]{O}$

7) Write a differential equation that is a mathematical model of the situation in which the time rate of change in the population of a certain organism is proportional to the product of the current population and the difference between the current population and the limiting factor of 100,000.

   Answer: $\frac{dP}{dt} = kP(100{,}000 - P)$

8) Suppose that $1000 is deposited in a savings account that pays 6% annual interest compounded continuously. At what rate (in dollars per year) is it earning interest after 5 years?

   Answer: $80.99 per year

9) Carbon extracted from an ancient skull recently unearthed contained only one-fifth as much radioactive carbon 14 as carbon extracted from present-day bone. How old is the skull?

Answer: approximately 13,235 years old

10) A bacteria population is increasing according to the natural growth formula and numbers 100 at 12 noon and 156 at 1 p.m. Write a formula giving P(t) after t hours.

Answer: $P(t) = 100e^{0.4447t}$

11) The water in a draining cylindrical tank is 12 feet deep at 1:00 p.m. At 3:00 p.m. it is 4 feet deep. When will the tank be empty?

Answer: at approximately 5:44 p.m. ($t \approx 4.73$)

9.2 Slope Fields and Euler's Method
Solve the problem.

1) An initial value problem is given below. Apply Euler's method to approximate this solution on the interval $[0, \frac{1}{2}]$. Use step size h = 0.1.

$\frac{dy}{dx} = 3y$, $y(0) = 1$

Answer: $x_0 = 0$, $y_0 = 1$
$x_1 = 0.1$, $y_1 = 1.3$
$x_2 = 0.2$, $y_2 = 1.69$
$x_3 = 0.3$, $y_3 = 2.197$
$x_4 = 0.4$, $y_4 = 2.8561$
$x_5 = 0.5$, $y_5 = 3.71293$

2) An initial value problem is given below. Apply Euler's method to approximate this solution on the interval $[0, \frac{1}{2}]$. Use step size h = 0.1.

$\frac{dy}{dx} = -(y + x)$, $y(0) = -1$

Answer: $x_0 = 0$, $y_0 = -1$
$x_1 = 0.1$, $y_1 = -0.9$
$x_2 = 0.2$, $y_2 = -0.82$
$x_3 = 0.3$, $y_3 = -0.758$
$x_4 = 0.4$, $y_4 = -0.7122$
$x_5 = 0.5$, $y_5 = -0.68098$

9.3 Separable Equations and Applications
Solve the problem.

1) Find a general solution (implicit if necessary, explicit if possible) of the differential equation $\frac{dy}{dx} = 2xy$.

Answer: $y = Ae^{x^2}$

2) Find a general solution (implicit if necessary, explicit if possible) of the differential equation $\frac{dy}{dx} = 4x^2(y-1)^3$.

Answer: $y = 1 + \dfrac{1}{\sqrt{-\frac{8}{3}x^3 + C}}$

3) Solve the initial value problem $\frac{dy}{dx} = y^3$, $y(0) = 1$.

Answer: $y = \dfrac{1}{\sqrt{-2x+1}}$

4) Use the method of linear differential equations to solve the initial value problem $\frac{dx}{dt} = 3x - 4$, $x(0) = 2$.

Answer: $x(t) = 2e^{3t} - \frac{4}{3}(e^{3t} - 1)$

5) Nyobia had a population of 3 million in 1985. Assume that this country's population is growing continuously at a 5% annual rate and that Nyobia absorbs 40,000 newcomers per year. What will its population be in the year 2015?

Answer: 16.23 million

6) How much would you need to invest per month – in effect, continuously – in an investment account that pays an annual interest rate of 9%, compounded continuously, in order for the account to be worth $100,000 after 20 years?

Answer: $148.53

7) Suppose Anytown, USA has a fixed population of 200,000. On March 1, 3000 people have the flu. On June 1, 6000 people have it. If the rate of increase of the number N(t) who have the flu is proportional to the number who don't have it, how many will have the disease on September 1?

Answer: approximately 8954 people

9.4 Linear Equations and Applications
Solve the problem.

1) Find the particular solution of the differential equation $\frac{dy}{dx} + y = 3$ with initial condition $y(0) = 0$.

Answer: $y = 3 - 3e^{-x}$

2) Find the particular solution of the differential equation $x\frac{dy}{dx} + 3y = 5x$ subject to the initial condition $y(1) = 1$.

Answer: $y(x) = \frac{5}{4}x - \dfrac{1}{4x^3}$

3) A tank contains 2000 L of a solution consisting of 50 kg of salt dissolved in water. Pure water is pumped into the tank at the rate of 10L/s, and the mixture (kept uniform by stirring) is pumped out at the same rate. How long will it be until only 5 kg of salt remain in the tank?

Answer: approximately 518 seconds

4) Suppose that a motorboat is moving at 30 ft/s when its motor suddenly quits, and that 5 seconds later the boat has slowed to 15 ft/s. Assume that the resistance it encounters while coasting is proportional to its velocity. How far will the boat coast in all?

Answer: approximately 216 feet

9.5 Population Models
Solve the problem.

1) Use partial fractions to solve the initial value problem; $\frac{dx}{dt} = 2x(10 - x)$, $x(0) = 5$.

   Answer: $x(t) = \frac{10e^{20t}}{1 + e^{20t}}$

2) Suppose that in 1900 the population of a certain country was 30 million and was growing at the rate of 600,00 people per year at that time. Suppose also that in 1950 its population was 60 million and was then growing at the rate of 900,000 people per year. Assume that this population satisfies the logistic equation. Determine both the limiting population M and the predicted population for the year 2000.

   Answer: The limiting population M is 150 million. The predicted population for the year 2000 is approximately 112,923,000.

3) The time rate of change of an alligator population P in a swamp is proportional to the square root of P. The swamp contained 9 alligators in 1990 and 25 alligators in 1995. When will there be 49 alligators in the swamp?

   Answer: in the year 2000

4) Suppose that an animal population is modeled by the differential equation $\frac{dP}{dt} = 0.002P(P-150)$ and that $P(0) = 200$. How long will it take for this population to double to 400 animals? Is this an example of a doomsday situation or an extinction situation? Why?

   Answer: It will take approximately 3.05 years for this population to double. This is an example of a doomsday situation because the initial size of 200 exceeds the threshold size of 150.

9.6 Linear Second Order Equations
Solve the problem.

1) Find general solutions of the differential equation $y'' - 5y' + 6y = 0$.

   Answer: $y(x) = c_1 e^{2x} + c_2 e^{3x}$

2) Find general solutions of the differential equation $6y'' - 11y' - 10y = 0$.

   Answer: $y(x) = c_1 e^{5x/2} + c_2 e^{-2x/3}$

3) Find general solutions of the differential equation $81y'' - 72y' + 16y = 0$.

   Answer: $y(x) = (c_1 + c_2 x)e^{4x/9}$

4) Find general solutions of the differential equation $y'' + 2y' + 5y = 0$.

   Answer: $y(x) = e^{-x}(c_1 \cos 2x + c_2 \sin 2x)$

5) Solve the initial value problem $y'' - 7y' + 12y = 0$; $y(0) = 5$, $y'(0) = 18$.

   Answer: $y(x) = 2e^{3x} + 3e^{4x}$

6) Solve the initial value problem: $y'' - 6y' + 9y = 0$; $y(0) = 4$, $y'(0) = 13$.

   Answer: $y(x) = (4 + x)e^{3x}$

7) Solve the initial value problem $y'' + 16y = 0$; $y(0) = 2$, $y'(0) = 4$.

   Answer: $y(x) = 2\cos 4x + \sin 4x$

8) Solve the initial value problem: $y'' + 6y' + 25y = 0$; $y(0) = 5$, $y'(0) = 1$.

   Answer: $y(x) = e^{-3x}(5\cos 4x + 4\sin 4x)$

9) The general solution of a homogeneous linear second-order differential equation with constant coefficients is $y(x) = c_1 e^{3x} + c_2 e^{4x}$. Find that equation.

   Answer: $y'' - 7y' + 12y = 0$

10) The general solution of a homogeneous linear second-order differential equation with constant coefficients is $y(x) = (c_1 + c_2 x)e^{-4x}$. Find that equation.

    Answer: $y'' + 8y' + 16y = 0$

11) The general solution of a homogeneous linear second-order differential equation with constant coefficients is $y(x) = c_1 \cos 4x + c_2 \sin 4x$. Find that equation.

    Answer: $y'' + 16y = 0$

12) The general solution of a homogeneous linear second-order differential equation with constant coefficients is $y(x) = e^{-3x}(c_1 \cos 2x + c_2 \sin 2x)$. Find that equation.

    Answer: $y'' + 6y' + 13y = 0$

9.7 Mechanical Vibrations
Solve the problem.

1) This problem concerns undamped free motion of a mass $m = 1$ on a spring with Hooke's (spring) constant $k = 25$. Suppose that the mass is set in motion with initial position $x(0) = 3$ and initial velocity $x'(0) = 30$. Write the position function of the mass in the form $x(t) = C\cos(\omega_0 t - \alpha)$.

   Answer: $x(t) = 3\cos(5t - 1.1071)$

2) This problem concerns damped free motion of a mass $m = 1$ that is attached both to a spring with Hooke's constant $k = 25$ and to a dashpot with damping constant $c = 6$. Suppose that the mass is set in motion with initial position $x(0) = 2$ and initial velocity $x'(0) = 4$. Find the position function of the mass in the form $x(t) = C\cos(\omega_0 t - \alpha)$. Determine whether the resulting motion is overdamped, critically damped, or underdamped. In the latter case, write the position function in the form $x(t) = Ce^{-pt}\cos(\omega_1 t - \alpha)$.

   Answer: $x(t) = 2\sqrt{5}e^{-3t}\cos(4t - 1.1071)$ (underdamped)

3) The equation $x'' + 16x = 28\cos 3t$; $x(0) = x'(0) = 0$, describes forced undamped motion of a mass on a spring. Express the position function $x(t)$ as the sum of two oscillations.

   Answer: $x(t) = -4\cos 4t + 4\cos 3t$

4) Find the steady periodic solution of the given differential equation $x'' + 2x' + 17x = 59\cos 3t$; $x(0) = x'(0) = 0$. Also find the transient solution.

   Answer: $x(t) = 8.5\cos 3t + 3\sin 3t$
   $x(t) = e^{-t}(-8.5\cos 4t - 4.375\sin 4t)$

5) Determine the period and frequency of the simple harmonic motion of a 1-kg mass on the end of a spring with spring constant 4 N/m.

Answer: period = π; frequency = 2

6) A weight of $\frac{1}{2}$ lb. is attached to a vertical spring with spring constant k = 1 lb./ft. The weight is pushed 2 ft. above its equilibrium position and released. Find (a) the position function of the weight, (b) the amplitude of the vibration, (c) the period of the vibration, and (d) the frequency of the vibration.

Answer: (a) x(t) = 2cos 8t (b) 2 (c) $\frac{\pi}{4}$ (d) $\frac{4}{\pi}$

7) A mass of 25 g is attached to a vertical spring with spring constant k = 3 dyne/cm. The surrounding medium has a damping constant of c = 10 dyne·sec/cm. The mass is pushed 5 cm above its equilibrium position and released. Find (a) the position function of the mass, (b) the period of the vibration, and (c) the frequency of the vibration.

Answer: (a) $x(t) = \frac{5}{2}\sqrt{6} e^{-t/8} \cos(\frac{\sqrt{2}}{5} t - .61548)$ (b) $\frac{10\pi}{\sqrt{2}}$ (c) $\frac{\sqrt{2}}{10\pi}$

8) A weight of 3 lb. attached to a vertical spring stretches the spring $\frac{1}{2}$ ft. While in its equilibrium position, the weight is struck to give it a downward initial velocity of 2 ft./sec. Assuming that there is no damping, find (a) the position function of the weight, (b) the amplitude of the vibration, (c) the period of the vibration, and (d) the frequency of the vibration.

Answer: (a) $x(t) = -\frac{1}{4}\sin 8t$ (b) $\frac{1}{4}$ (c) $\frac{\pi}{4}$ (d) $\frac{4}{\pi}$

Ch. 10 Polar Coordinates and Parametric Curves
10.1 Analytic Geometry and the Conic Sections
Solve the problem.

1) Write the equation of the line that goes through the point (2, −3) and is parallel to the line with the equation $x + 3y = 7$.

   Answer: $y = -\frac{1}{3}x - \frac{7}{3}$

2) Write the equation of the line that goes through the point (−4, 1) and is perpendicular to the line with the equation $5x - 3y = 1$.

   Answer: $y = -\frac{3}{5}x - \frac{7}{5}$

3) Write the equation of the line that is tangent to the circle $x^2 + y^2 = 100$ at the point (−6, −8).

   Answer: $y = -\frac{3}{4}x - \frac{25}{2}$

4) Write the equation of the line that is tangent to the curve $y^2 = x + 7$ at the point (−3, 2).

   Answer: $y = -\frac{1}{4}x + \frac{11}{4}$

5) Write the equation of the line that is perpendicular to the curve $x^2 + 3y^2 = 12$ at the point (3, 1).

   Answer: $y = x - 2$

6) Write the equation of the line that is the perpendicular bisector of the line segment with endpoints (3, 4) and (−5, −1).

   Answer: $y = -\frac{8}{5}x - \frac{1}{10}$

7) Find the center and radius of the circle described in the equation $x^2 + 4x + y^2 = 16$.

   Answer: center at (−2, 0); $r = 2\sqrt{5}$

8) Find the center and radius of the circle described in the equation $x^2 + y^2 - 10x + 12y = 3$.

   Answer: center at (5, −6); $r = 8$

9) Find the center and radius of the circle described in the equation $4x^2 + 4y^2 - 4y = 35$.

   Answer: center at $(0, \frac{1}{2})$; $r = 3$

10) Find the center and radius of the circle described in the equation $2x^2 + 2y^2 - 6x + 2y = 3$.

    Answer: center at $(\frac{3}{2}, -\frac{1}{2})$; $r = 2$

11) Find the center and radius of the circle described in the equation $400x^2 + 400y^2 - 640x - 1000y = 2719$.

    Answer: center at $(\frac{4}{5}, \frac{5}{4})$; $r = 3$

Calculus 197

12) Write an equation for the circle with center (-2, -3) that passes through the point (5, 6).

   Answer: $x^2 + 4x + y^2 + 6y = 117$.

13) Write an equation for the circle with center (-3, -5) that is tangent to the line $y = \frac{3}{4}x + 1$.

   Answer: $x^2 + 6x + y^2 + 10y = -25$.

14) Write an equation for the circle that passes through the points (1, -1), (-5, 7), and (-6, 0).

   Answer: $x^2 + 4x + y^2 - 6y = 12$.

15) Write an equation for the set of all points P(x, y) such that P(x, y) is equally distant from the two points (1, 5) and (7, 3).

   Answer: $y = 3x - 8$

16) Write an equation for the set of all points P(x, y) such that P(x, y) is half the distance from the point (1, 2) as from the point (-2, -4).

   Answer: $(x - 2)^2 + (y - 4)^2 = 20$

17) Write an equation for the set of all points P(x, y) such that P(x, y) is 3 times the distance from the point (-8, 4) as from the point (4, 8).

   Answer: $(x - \frac{11}{2})^2 + (y - \frac{17}{2})^2 = \frac{45}{2}$

18) Write an equation for the set of all points P(x, y) such that P(x, y) is an equal distance from the line x = -5 and the point (5, 0).

   Answer: $y^2 = 20x$

19) Find all the lines through the point (3, 2) that are tangent to the parabola $y = x^2$.

   Answer: $y = (6 + 2\sqrt{7})x - 16 - 6\sqrt{7} \approx 11.292x - 31.875$
   $y = (6 - 2\sqrt{7})x - 16 + 6\sqrt{7} \approx 0.708x - 0.125$

20) Find all lines that are normal to the curve xy = 3 and are simultaneously parallel to the line y = 3x.

   Answer: $y = 3x - 8$
   $y = 3x + 8$

10.2 Polar Coordinates
Solve the problem.

1) Plot the point with the polar coordinates $(2, \frac{\pi}{4})$. Then find the rectangular coordinates.

Answer: $(\sqrt{2}, \sqrt{2})$

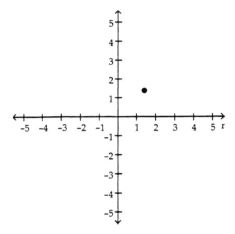

2) Plot the point with the polar coordinates $(-1, \frac{2\pi}{3})$. Then find the rectangular coordinates.

Answer: $(\frac{1}{2}, -\frac{\sqrt{3}}{2})$

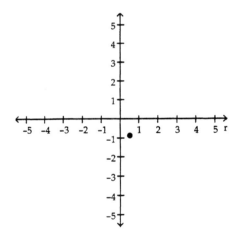

3) Plot the point with the polar coordinates $(3, -\frac{\pi}{3})$. Then find the rectangular coordinates.

Answer: $(\frac{3}{2}, -\frac{3\sqrt{3}}{2})$

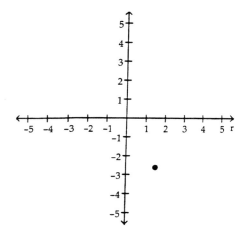

4) Plot the point with the polar coordinates $(2, \frac{3\pi}{2})$. Then find the rectangular coordinates.

Answer: $(0, -2)$

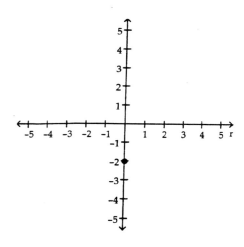

5) Plot the point with the polar coordinates $(5, -\frac{\pi}{4})$. Then find the rectangular coordinates.

Answer: $(\frac{5\sqrt{2}}{2}, -\frac{5\sqrt{2}}{2})$

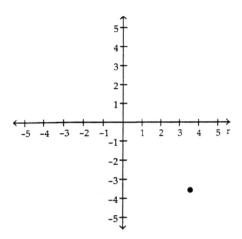

6) Plot the point with the polar coordinates $(-3, -\frac{7\pi}{6})$. Then find the rectangular coordinates.

Answer: $(\frac{3\sqrt{3}}{2}, -\frac{3}{2})$

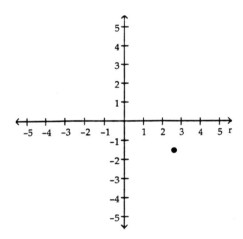

7) Plot the point with the polar coordinates $(1, \frac{5\pi}{6})$. Then find the rectangular coordinates.

Answer: $(-\frac{\sqrt{3}}{2}, \frac{1}{2})$

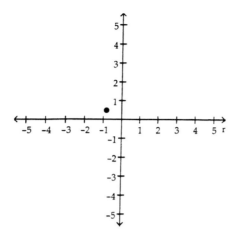

8) Find two polar coordinate representations, one with r > 0 and the other with r < 0, for the point with the rectangular coordinates (–1, 1).

Answer: $(\sqrt{2}, \frac{3\pi}{4})$ and $(-\sqrt{2}, \frac{7\pi}{4})$

9) Find two polar coordinate representations, one with r > 0 and the other with r < 0, for the point with the rectangular coordinates $(\sqrt{3}, 1)$.

Answer: $(2, \frac{\pi}{6})$ and $(-2, \frac{7\pi}{6})$

10) Find two polar coordinate representations, one with r > 0 and the other with r < 0, for the point with the rectangular coordinates (–2, –2).

Answer: $(2\sqrt{2}, \frac{5\pi}{4})$ and $(-2\sqrt{2}, \frac{\pi}{4})$

11) Find two polar coordinate representations, one with r > 0 and the other with r < 0, for the point with the rectangular coordinates $(3, -\sqrt{3})$.

Answer: $(2\sqrt{3}, \frac{11\pi}{6})$ and $(-2\sqrt{3}, \frac{5\pi}{6})$

12) Express the rectangular equation x = 5 in polar form.

Answer: $r = 5\sec\theta$

13) Express the rectangular equation x = 2y in polar form.

Answer: $\theta = \arctan(\frac{1}{2})$

14) Express the rectangular equation y = 2 in polar form.

Answer: $r = 2\csc\theta$

15) Express the rectangular equation $x^2 + y^2 = 16$ in polar form.

Answer: $r = 4$

16) Express the rectangular equation $xy = 2$ in polar form.

Answer: $r^2 = \dfrac{2}{(\cos\theta)(\sin\theta)}$

17) Express the rectangular equation $x^2 - y^2 = 1$ in polar form.

Answer: $r^2 = \dfrac{1}{\cos 2\theta}$

18) Express the rectangular equation $x = y^2$ in polar form.

Answer: $r = (\cot\theta)(\csc\theta)$

19) Express the rectangular equation $x + y = 9$ in polar form.

Answer: $r = \dfrac{9}{\cos\theta + \sin\theta}$

20) Express the polar equation $r = 7$ in rectangular form.

Answer: $x^2 + y^2 = 49$

21) Express the polar equation $\theta = \dfrac{7\pi}{4}$ in rectangular form.

Answer: $y = -x$

22) Express the polar equation $r = -2\cos\theta$ in rectangular form.

Answer: $x^2 + 2x + y^2 = 0$

23) Express the polar equation $r = 3\sin 2\theta$ in rectangular form.

Answer: $(x^2 + y^2)^3 = 36x^2y^2$

24) Express the polar equation $r = 1 - \cos 2\theta$ in rectangular form.

Answer: $(x^2 + y^2)^3 = 4y^4$

25) Express the polar equation $r = 1 - \cos 2\theta$ in rectangular form.

Answer: $(x^2 - x + y^2)^2 = 9(x^2 + y^2)$

26) Express the polar equation $r = 5\csc\theta$ in rectangular form.

Answer: $y = 5$

27) Express the polar equation $r^2 = 2\cos 2\theta$ in rectangular form.

Answer: $(x^2 + y^2)^2 = 2x^2 - 2y^2$

28) Write an equation in both rectangular and polar form for the vertical line through $(-3, 0)$.

Answer: $x = -3$
$r = -3\sec\theta$

29) Write an equation in both rectangular and polar form for the horizontal line through (2, 5).

Answer: $y = 5$
$r = 5\csc\theta$

30) Write an equation in both rectangular and polar form for the line with slope –1 through (3, –1).

Answer: $x + y = 2$
$r = \dfrac{2}{\cos\theta + \sin\theta}$

31) Write an equation in both rectangular and polar form for the line through (1, 4) and (5, 8).

Answer: $y = x + 2$
$r = \dfrac{3}{\sin\theta - \cos\theta}$

32) Write an equation in both rectangular and polar form for the circle with center (0, –7) that passes through the origin.

Answer: $x^2 + y^2 + 14y = 0$
$r = -14\sin\theta$

33) Write an equation in both rectangular and polar form for the circle with center (8, 6) and radius 10.

Answer: $x^2 - 16x + y^2 - 12y = 0$
$r = 16\cos\theta + 12\sin\theta$

34) Write an equation in both rectangular and polar form for the circle with center (8, 6) that passes through (2, –6).

Answer: $x^2 - 12x + y^2 + 6y = -20$
$r^2 = 12r\cos\theta + 6r\sin\theta - 20$

35) Match the polar equations to the their rectangular graphs by first transforming the equations to rectangular form.

1) $r = -6\cos\theta$
2) $r = 3\cos\theta + 3\sin\theta$
3) $r = -3\cos\theta + 4\sin\theta$
4) $r = 5\cos\theta - 12\sin\theta$

A)

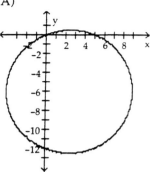

B)

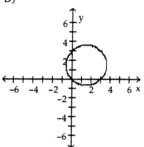

C)

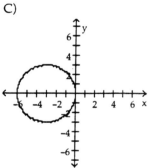

D)

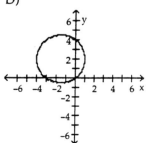

Answer: 1C, 2B, 3D, 4A

36) Match the polar equations to the their rectangular graphs by first transforming the equations to rectangular form.

1) $r = 6 + 4\cos\theta$
2) $r = 6 + 6\cos\theta$
3) $r = 5 + 7\cos\theta$
4) $r = 1 + 5\cos\theta$

A)
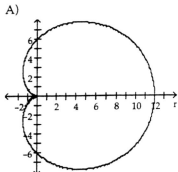

B)
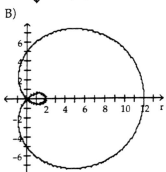

C)
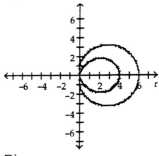

D)
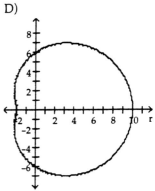

Answer: 1D, 2A, 3B, 4C

37) Sketch the graph of the polar equation $r = 5\sin\theta$ (circle). Indicate any symmetries around either coordinate axis or the origin.

Answer: symmetric about the y-axis

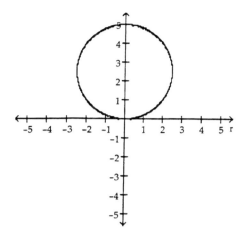

38) Sketch the graph of the polar equation $r = 3\sin\theta + 3\cos\theta 3\sin\theta + 3\cos\theta$ (circle). Indicate any symmetries around either coordinate axis or the origin.

Answer: no symmetries

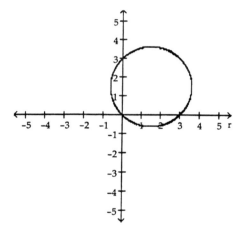

39) Sketch the graph of the polar equation r = 1 - cos θ (cardioid). Indicate any symmetries around either coordinate axis or the origin.

Answer: symmetric about the x-axis

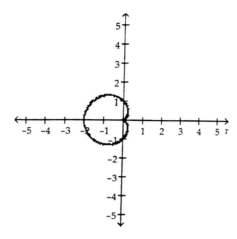

40) Sketch the graph of the polar equation r = 1 + sin θ (cardioid). Indicate any symmetries around either coordinate axis or the origin.

Answer: symmetric about the y-axis

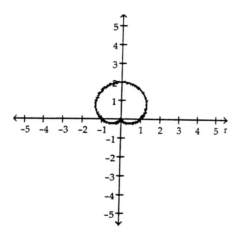

41) Sketch the graph of the polar equation r = 1 + 2sin θ (limacon). Indicate any symmetries around either coordinate axis or the origin.

Answer: symmetric about the y-axis

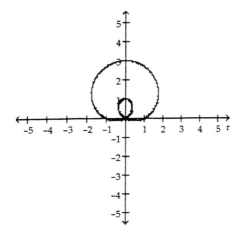

42) Sketch the graph of the polar equation r = 2 + cos θ (limacon). Indicate any symmetries around either coordinate axis or the origin.

Answer: symmetric about the x-axis

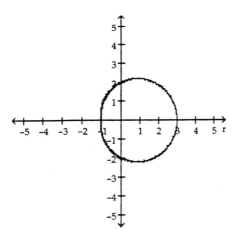

Calculus 209

43) Sketch the graph of the polar equation $r^2 = 9\sin 2\theta$ (lemniscate). Indicate any symmetries around either coordinate axis or the origin.

Answer: symmetric about the origin

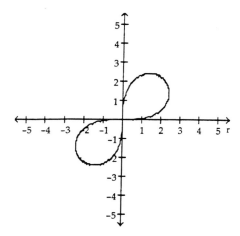

44) Sketch the graph of the polar equation $r^2 = 9\cos 2\theta$ (lemniscate). Indicate any symmetries around either coordinate axis or the origin.

Answer: symmetric about both axes and the origin

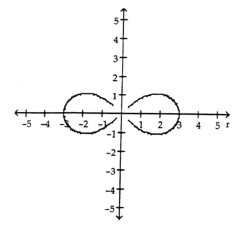

45) Sketch the graph of the polar equation r = 3sin 2θ (four-leaved rose). Indicate any symmetries around either coordinate axis or the origin.

Answer: symmetric about both axes and the origin

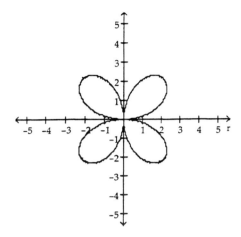

46) Sketch the graph of the polar equation r = 4sin 3θ (three-leaved rose). Indicate any symmetries around either coordinate axis or the origin.

Answer: symmetric about the y-axis

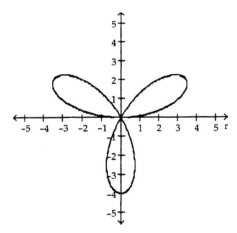

47) Sketch the graph of the polar equation r = 2cos 3θ (three-leaved rose). Indicate any symmetries around either coordinate axis or the origin.

Answer: symmetric about the x-axis

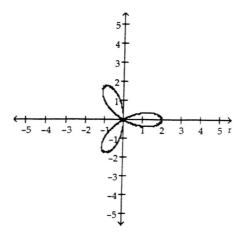

48) Sketch the graph of the polar equation r = 2θ (spiral of Archimedes). Indicate any symmetries around either coordinate axis or the origin.

Answer: no symmetries

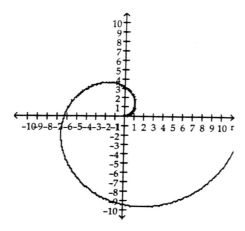

49) Sketch the graph of the polar equation r = 5sin 5θ (five-leaved rose). Indicate any symmetries around either coordinate axis or the origin.

Answer: symmetric about the y-axis

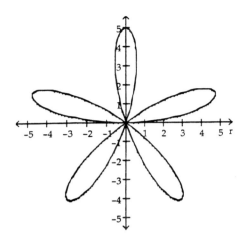

50) Sketch the graph of the polar equation r^2 = 9sin θ (figure eight). Indicate any symmetries around either coordinate axis or the origin.

Answer: symmetric about both axes and the origin

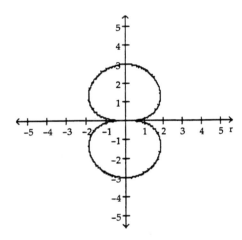

51) Find all points of intersection of the polar equations r = 2 and r = 2cos θ.

Answer: (2, 0)

52) Find all points of intersection of the polar equations r = 3sin θ and r = 3cos 2θ.

Answer: (0, 0), $(\frac{3}{2}, \frac{\pi}{6})$, $(\frac{3}{2}, \frac{5\pi}{6})$, $(3, \frac{\pi}{2})$

53) Find all points of intersection of the polar equations r = 1 − sin θ and r^2 = 4sin θ.

Answer: (0, 0), $(2, \frac{3\pi}{2})$, and the two points where r = $2(\sqrt{2} - 1)$ and θ or (π − θ) = arcsin $(3 - 2\sqrt{2})$

Calculus 213

10.3 Area Computations in Polar Coordinates
Solve the problem.

1) Sketch the plane region bounded by the polar curve $r = 2\theta$, $0 \le \theta \le \pi$.

   Answer:

   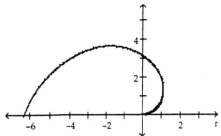

2) Sketch the plane region bounded by the polar curve $r = 1/\theta$, $0 \le \theta \le 2\pi$.

   Answer:

   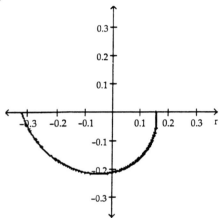

3) Sketch the plane region bounded by the polar curve $r = e^{-\theta}$, $0 \le \theta \le \pi$.

   Answer:

   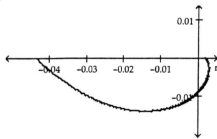

4) Find the area bounded by one loop of the polar equation $r = 4 \cos 2\theta$.

   Answer: 2π

5) Find the area enclosed by the small loop of the polar equation $r = -2 + 4 \cos \theta$.

   Answer: $-6\sqrt{3} + 4\pi$

6) Find the area bounded by the cardioid with polar equation $r = 1 + \sin\theta$.

   Answer: $\dfrac{3\pi}{2}$

7) Find the area bounded by one loop of the graph of the polar equation $r = 2\sin 3\theta$.

   Answer: $\dfrac{\pi}{3}$

8) Find the area enclosed by the graph of the polar equation $r = \sin\theta + \cos\theta$.

   Answer: $\dfrac{\pi}{2}$

9) Find the area enclosed of the graph of the polar equation $r = 2 - \sin\theta$.

   Answer: $\dfrac{9\pi}{2}$

10) Find the area bound by one loop of the graph of the polar equation $r = \sin\theta \cos\theta$.

    Answer: $\dfrac{\pi}{32}$

11) Find the area enclosed by the graph of the polar equation $r = 4 + 2\sin\theta$.

    Answer: 18π

12) Find the area enclosed by one loop of the graph of the polar equation $r^2 = 4\sin 2\theta$.

    Answer: 2

13) Find the area enclosed by the graph of the polar equation $r = 1 - \sin\theta$.

    Answer: $\dfrac{2}{3}\pi$

14) Find the area enclosed by the graph of the polar equation $r = 6\cos\theta$.

    Answer: 9π

15) Find the area enclosed by the graph of the polar equation $r = 4\sin 3\theta$.

    Answer: 4π

16) Find the area enclosed by the graph of the polar equation $r = 2 - \cos\theta$.

    Answer: $\dfrac{9\pi}{2}$

17) Find the area of the region bounded by the curve $r = \theta$, $\pi \leq \theta \leq \dfrac{5}{2}\pi$, and the rays $\theta = \pi$ and $\theta = \dfrac{5}{2}\pi$.

    Answer: $\dfrac{39}{16}\pi^3$

18) Find the length of the equiangular spiral $r = e^\theta$ for $\theta \leq 0$.

    Answer: $\sqrt{2}$

19) Find the total area enclosed by the graph of the polar equation $r = 1 + \cos 2\theta$.

Answer: $\dfrac{3\pi}{2}$

20) Find the total area enclosed by the graph of the polar equation $r = \sin^2 \theta$.

Answer: $\dfrac{3\pi}{8}$

21) Find the total area enclosed by the graph of the polar equation $r = \sin 4\theta$.

Answer: $\dfrac{\pi}{2}$

22) Find the total area enclosed by the graph of the polar equation $r = \sin 6\theta$.

Answer: $\dfrac{\pi}{2}$

23) Find the total area enclosed by the graph of the polar equation $r = \sin 8\theta$.

Answer: $\dfrac{\pi}{2}$

24) Find area of the region inside both $r = 6\cos \theta$ and $r = 3$.

Answer: $\dfrac{27\sqrt{3} - \pi}{6}$

25) Find area of the region inside $r = 3 + \cos \theta$ and outside $r = 3$.

Answer: $\pi + 6$

26) Find area of the region inside $r^2 = 2\sin 2\theta$ and outside $r = 1$.

Answer: $\sqrt{3} - \dfrac{\pi}{3}$

27) Find area of the region inside the large loop and outside the small loop of $r = 2 - 4\sin \theta$.

Answer: $4\pi + 12\sqrt{3}$

28) Find area of the region inside the figure-eight curve $r^2 = 4\sin \theta$ and outside $r = 1 - \sin \theta$.

Answer: ≈ 3.729

29) Find area of the region inside the $r = 3 + 3\sin \theta$ and outside $r = 3$.

Answer: $9(2 + \dfrac{\pi}{4})$

30) Find area of the region that lies interior to all three circles $r = 2$, $r = 4\cos \theta$, and $r = \sin \theta$.

Answer: $\dfrac{5\pi}{3} - 2\sqrt{3}$

10.4 Parametric Curves
Solve the problem.

1) Eliminate the parameter in $x = t + 2$ and $y = 2t + 5$. Then sketch the curve.

   Answer: $y = 2x + 1$

   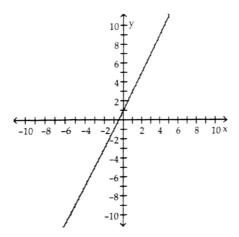

2) Eliminate the parameter in $x = t^2$ and $y = t^5$. Then sketch the curve.

   Answer: $y^2 = x^5$

   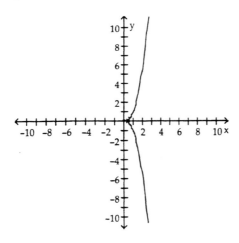

3) Eliminate the parameter in $x = \sqrt{t}$ and $y = 2t - 3$. Then sketch the curve.

Answer: $y = 2x^2 - 3$, $x > 0$

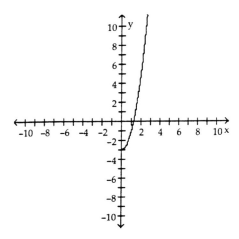

4) Eliminate the parameter in $x = e^t$ and $y = 2e^{3t}$. Then sketch the curve.

Answer: $y = 2x^3$, $x > 0$

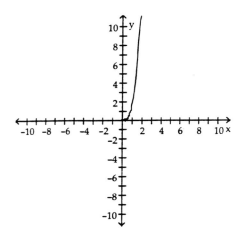

5) Eliminate the parameter in x = 4sin t and y = 7cos t. Then sketch the curve.

Answer: $(\frac{x}{4})^2 + (\frac{y}{7})^2 = 1$

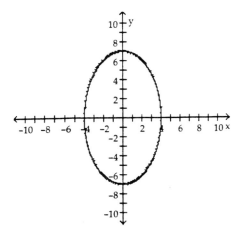

6) Eliminate the parameter in x = 3cosh t and y = 2sinh t. Then sketch the curve.

Answer: $(\frac{x}{3})^2 + (\frac{y}{2})^2 = 1$, x > 0

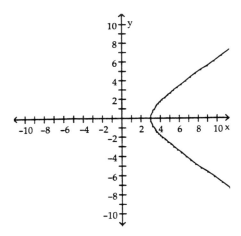

7) Eliminate the parameter in $x = \csc t$ and $y = \cot t$. Then sketch the curve.

   Answer: $x^2 - y^2 = 1$

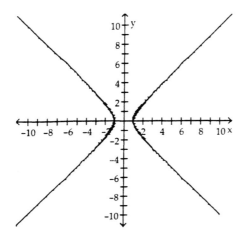

8) Eliminate the parameter in $x = \cos \pi t$ and $y = \sin \pi t$, $0 \le t \le 2$, and sketch the curve. Then describe the motion of the point $(x(t), y(t))$ in the given interval.

   Answer: $x^2 + y^2 = 1$

   The point starts at $(1, 0)$ and moves in a counter-clockwise direction around the unit circle, stopping at $(1, 0)$.

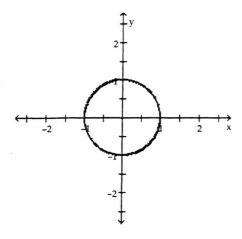

9) Eliminate the parameter in x = 2 + 3cos t and y = 4 + 5sin t, 0 ≤ t ≤ 2π, and sketch the curve. Then describe the motion of the point (x(t), y(t)) in the given interval.

Answer: $(\frac{x-2}{3})^2 + (\frac{y-4}{5})^2 = 1$

The point starts at (5, 4) and moves in a counter-clockwise direction around the ellipse, stopping at (5, 4).

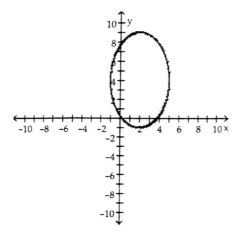

10) Write the equation of the line tangent to the parametric curve x = 7t² - 5, y = 4t³ + 1 at the point corresponding to t = 1. Then calculate d²y/dx² to determine whether the curve is concave upward or concave downward at this point.

Answer: $y - 5 = \frac{6}{7}(x - 2)$

$d^2y/dx^2 = \frac{3}{49}t^{-1}$; concave upward

11) Write the equation of the line tangent to the parametric curve x = sin³ t, y = cos³ t at the point corresponding to t = π/4. Then calculate d²y/dx² to determine whether the curve is concave upward or concave downward at this point.

Answer: $y = -x + \frac{\sqrt{2}}{2}$

$d^2y/dx^2 = \frac{(\csc^4 t)(\sec t)}{3}$; concave upward

12) Write the equation of the line tangent to the parametric curve x = tcos t, y = tsin t at the point corresponding to t = 2π. Then calculate d²y/dx² to determine whether the curve is concave upward or concave downward at this point.

Answer: $y = 2\pi x - 4\pi^2$

$d^2y/dx^2 = \frac{t^2 + 2}{(\cos t - t\sin t)^3}$; concave upward

13) Write the equation of the line tangent to the parametric curve $x = e^{-t}$, $y = e^t$ at the point corresponding to $t = 0$. Then calculate d^2y/dx^2 to determine whether the curve is concave upward or concave downward at this point.

Answer: $y - 1 = -(x - 1)$
$d^2y/dx^2 = 2e^{3t}$; concave upward

14) Given $r = \exp(\theta\frac{\sqrt{3}}{3})$, find the angle ψ between the radius OP and the tangent line at the point P that corresponds to $\theta = \frac{\pi}{4}$.

Answer: $\psi = \frac{\pi}{3}$

15) Given $r = 2/\theta$, find the angle ψ between the radius OP and the tangent line at the point P that corresponds to $\theta = 1$.

Answer: $\psi = \frac{3\pi}{4}$

16) Given $r = \sin 2\theta$, find the angle ψ between the radius OP and the tangent line at the point P that corresponds to $\theta = \frac{\pi}{4}$.

Answer: $\psi = \frac{\pi}{2}$

17) Given $r = 1 - \sin \theta$, find the angle ψ between the radius OP and the tangent line at the point P that corresponds to $\theta = \frac{\pi}{6}$.

Answer: $\psi = \frac{5\pi}{6}$

18) Find the points on the parametric curve $x = t^2$, $y = t^3 - 12t$ where the tangent line is horizontal. Then find the slope of each tangent line at any point where the curve intersects the x-axis.

Answer: There are horizontal tangents at the points (4, 16) and (4, -16). There is a vertical tangent at the x-intercept (0, 0). There is no tangent line at the other x-intercept, (12, 0), because the curve crosses itself with two different slopes there, namely the slopes $\pm 2\sqrt{3}$.

19) Find the points on the parametric curve $r = 1 - \cos \theta$ where the tangent line is horizontal. Then find the slope of each tangent line at any point where the curve intersects the x-axis.

Answer: There are horizontal tangents at the points corresponding to $\theta = \pm \frac{2\pi}{3}$. The corresponding value of r is $\frac{3}{2}$, and the rectangular coordinates of these two points are $(-\frac{3}{4}, \pm \frac{3\sqrt{3}}{4})$. There is a vertical tangent at the x-intercept (-2, 0). (If a line tangent to the curve C at the point P is simply a line through P that approximates the curve's shape very, very, well at and near P, then there is a horizontal tangent line at (0, 0).)

20) Investigate the curve $x = y^3 + 3y^2 - 1$, and construct a sketch that shows all the critical points and inflection points on it.

Answer: Critical points are at $(-1,0)$ and $(3,-2)$. There is an inflection point at $(1, -1)$.

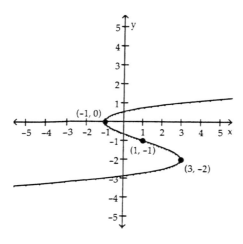

10.5 Integral Computations with Parametric Curves
Solve the problem.

1) The region bounded by the x-axis and the parametric curve $x = t^3$, $y = 1 - t^6$, $-1 \le t \le 1$, is rotated around the x-axis, Find the volume swept out.

   Answer: $\dfrac{16\pi}{15}$

2) The region R is bounded by the x-axis and the trajectory of the parametric curve $x = 2\cos t$, $y = \sin t$, $0 \le t \le \pi$. Find the volume obtained by rotation of R around the x-axis.

   Answer: $\dfrac{8\pi}{3}$

3) The region bounded by the x-axis and the parametrized curve $x = \cos t$, $y = e^t$, $0 \le t \le \pi$, is rotated around the x-axis, Find the volume swept out.

   Answer: $\dfrac{\pi(e^{2\pi} + 1)}{5} \approx 337.088$

4) The loop of the folium of Descartes $x^3 + y^3 = 3xy$ has the smooth parameterization $x = \dfrac{3t}{t^3 + 1}$, $y = \dfrac{3t^2}{t^3 + 1}$, $0 \le t \le +\infty$. Write and simplify the integral that gives the area enclosed by this loop.

   Answer: $\dfrac{3}{2}$

5) Find the area of the region that lies between the parametric equation $x = t^3$, $y = 5t^2 + 3$, $-1 \le t \le 1$ and the x-axis.

   Answer: 12

6) Find the area of the region that lies between the parametric equation $x = e^{4t}$, $y = e^{-t}$, $0 \le t \le \ln 3$ and the x-axis.

   Answer: $\dfrac{104}{3}$

7) Find the area of the region that lies between the parametric equation $x = \cos^2 t$, $y = \sin t$, $0 \le t \le \frac{\pi}{2}$ and the x-axis.

   Answer: $\frac{2}{3}$

8) Find the volume obtained by revolving around the x-axis the region described by the parametric equation $x = t^3$, $y = 5t^2 + 3$, $-1 \le t \le 1$.

   Answer: $\frac{528\pi}{7} \approx 236.9658$

9) Find the volume obtained by revolving around the x-axis the region described by the parametric equation $x = e^{4t}$, $y = e^{-t}$, $0 \le t \le \ln 3$.

   Answer: 16π

10) Find the volume obtained by revolving around the x-axis the region described by the parametric equation $x = \cos^2 t$, $y = \sin t$, $0 \le t \le \frac{\pi}{2}$.

    Answer: $\frac{\pi}{2}$

11) Find the arc length of the curve given by $x = 3t$, $y = \frac{2}{3}t^{3/2}$, $16 \le t \le 27$.

    Answer: $\frac{182}{3}$

12) Find the arc length of the curve given by $x = \sin t - \cos t$, $y = \sin t + \cos t$, $\frac{\pi}{1} \le t \le \frac{3\pi}{4}$.

    Answer: $\frac{5\sqrt{2}\pi}{12}$

13) Find the arc length of the curve given by $x = 2e^t \sin t$, $y = 2e^t \cos t$, $0 \le t \le \frac{\pi}{2}$.

    Answer: $2\sqrt{2}(e^{\pi/2}-1) \approx 10.7777$

14) Find the arc length of the curve given by $x = 2e^t \sin t$, $y = 2e^t \cos t$, $0 \le t \le \frac{\pi}{2}$.

    Answer: $2\sqrt{2}(e^{\pi/2}-1) \approx 10.7777$

15) Find the area of the surface of revolution generated by revolving the curve $x = 4 + t$, $y = \sqrt{t}$, $0 \le t \le 5$ around the x-axis.

    Answer: $\frac{\pi}{6}(21\sqrt{21} - 1) \approx 49.8645$

16) Find the area of the surface of revolution generated by revolving the curve $x = t^3$, $y = 5t - 1$, $-1 \le t \le 1$ around the y-axis.

    Answer: $\frac{2\pi}{27}(34\sqrt{34} - 125) \approx 17.0466$

17) Find the area of the surface of revolution generated by revolving the curve r = 3sin θ, 0 ≤ θ ≤ π around the x-axis.

   Answer: $9\pi^2 \approx 88.8264$

18) Find the area of the surface of revolution generated by revolving the curve r = 5sin θ, $0 \le t \le \frac{\pi}{2}$ around the y-axis.

   Answer: $25\pi \approx 78.5398$

19) The circle $x^2 + (y - b)^2 = a^2$ with radius a < b and center (0, b) can be parametrized by x = acos t, y = b + a sin t, 0 ≤ t ≤ 2π. Find the surface area of the torus obtained by revolving this circle around the x-axis.

   Answer: $4ab\pi^2$

20) The figure shows the graph of the parametric curve $x = t^2\sqrt{2}$, $y = 2t - \frac{1}{2}t^3$. The loop is bounded by the part of the curve fo which −2 ≤ t ≤ 2. Find its area.

   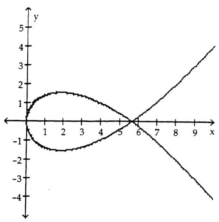

   Answer: $\frac{128\sqrt{2}}{15} \approx 12.0680$

10.6 Conic Sections and Applications
Solve the problem.

1) Find the equation and sketch the graph of the parabola with vertex V(0, 0) and focus F(2, 0).

   Answer: $y^2 = 8x$

   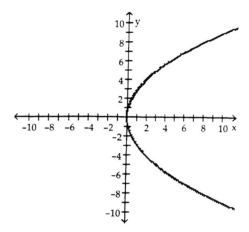

2) Find the equation and sketch the graph of the parabola with vertex V(1, 6) and focus F(1, 4).

   Answer: $(x - 1)^2 = 8(y - 6)$

   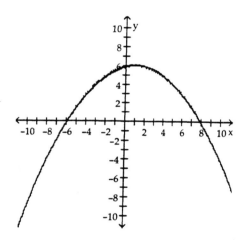

3) Find the equation and sketch the graph of the parabola with vertex V(-2, -2) and focus F(-3, -2).

Answer: $(y + 2)^2 = -4(x + 2)$

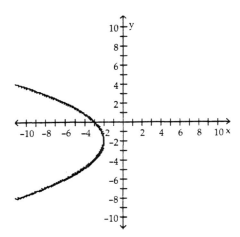

4) Find the equation and sketch the graph of the parabola with focus F(2, 1) and directrix x = -2.

Answer: $(y - 1)^2 = 8x$

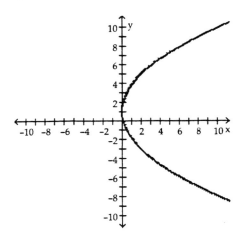

5) Find the equation and sketch the graph of the parabola with focus F(0, -5) and directrix x = 0.

Answer: $x^2 = -10(y + \frac{5}{2})$

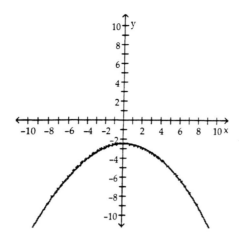

6) Find the equation and sketch the graph of the parabola with focus F(5, -2) and directrix x = 7.

Answer: $(y + 2)^2 = -8(x - 6)$

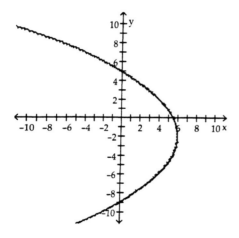

7) Find the equation and sketch the graph of the parabola with focus F(0, 0) and directrix x = −4.

Answer: $x^2 = -16(y + 2)$

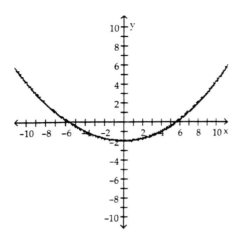

8) Sketch the parabola with the equation $y^2 = 8x$. Show and label its vertex, focus, axis, and directrix.

Answer: V(0, 0); F(2, 0); axis y = 0; directrix x = −2

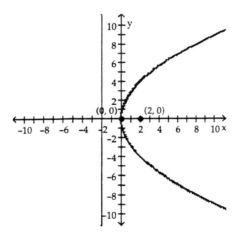

9) Sketch the parabola with the equation $x^2 = -12y$. Show and label its vertex, focus, axis, and directrix.

Answer: V(0, 0); F(0, -3); axis x = 0; directrix y = 3

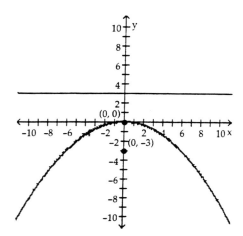

10) Sketch the parabola with the equation $x^2 + 4x - 4y = 0$. Show and label its vertex, focus, axis, and directrix.

Answer: V(-2, -1); F(-2, 0); axis x = -2; directrix y = -2

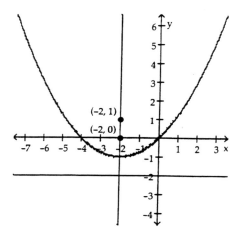

11) Sketch the parabola with the equation $y^2 - 2x + 8y + 18 = 0$. Show and label its vertex, focus, axis, and directrix.

Answer: $V(1, -4)$; $F(\frac{2}{3}, -4)$; axis $y = -4$; directrix $y = -\frac{1}{2}$

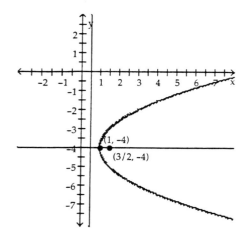

12) Sketch the parabola with the equation $4x^2 - 4x + 8y + 21 = 0$. Show and label its vertex, focus, axis, and directrix.

Answer: $V(\frac{1}{2}, -\frac{5}{2})$; $F(\frac{1}{2}, -3)$; axis $x = \frac{1}{2}$; directrix $y = -2$

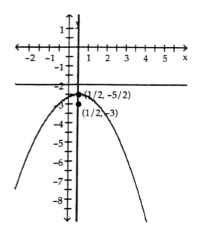

13) Find an equation of the ellipse with vertices $(\pm 3, 0)$ and $(0, \pm 6)$.

Answer: $(\frac{x}{5})^2 + (\frac{y}{6})^2 = 1$

14) Find an equation of the ellipse with foci $(0, \pm 12)$ and major semiaxis 13.

Answer: $(\frac{x}{5})^2 + (\frac{y}{13})^2 = 1$

15) Find an equation of the ellipse with center $(0, 0)$, vertical major axis 14, and minor axis 10.

Answer: $\frac{x^2}{25} + \frac{y^2}{49} = 1$

16) Find an equation of the ellipse with foci $(\pm 6, 0)$ and eccentricity $\frac{3}{4}$.

Answer: $\dfrac{x^2}{64} + \dfrac{y^2}{28} = 1$

17) Find an equation of the ellipse with center $(0, 0)$, horizontal major axis 18, and eccentricity $\frac{1}{3}$.

Answer: $\dfrac{x^2}{81} + \dfrac{y^2}{72} = 1$

18) Find an equation of the ellipse with foci $(\pm 3, 0)$ and directrices $x = \pm 12$.

Answer: $\dfrac{x^2}{36} + \dfrac{y^2}{27} = 1$

19) Find an equation of the ellipse with center $(-2, 1)$, horizontal major axis 10, and eccentricity $\frac{2}{5}$.

Answer: $\dfrac{(x+2)^2}{25} + \dfrac{(y-1)^2}{21} = 1$

20) Find an equation of the ellipse with foci $(10, 2)$, and $(-6, 2)$ and major axis 20.

Answer: $\dfrac{(x-2)^2}{100} + \dfrac{(y-2)^2}{36} = 1$

21) Find an equation of the ellipse with foci $(-5, 0)$, and $(-5, 6)$ and minor axis 12.

Answer: $\dfrac{(x+5)^2}{36} + \dfrac{(y+3)^2}{45} = 1$

22) Find an equation of the ellipse with foci $(-2, -4)$, and $(8, -4)$ and eccentricity $\frac{1}{2}$.

Answer: $\dfrac{(x-3)^2}{100} + \dfrac{(y+4)^2}{75} = 1$

23) Sketch the graph of the equation $4x^2 + 9y^2 = 324$. Indicate the center, foci, and lengths of the axes.

Answer: center $(0, 0)$, foci $(\pm 3\sqrt{5}, 0)$, major axis 18, minor axis 12

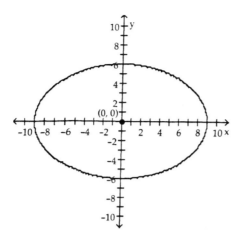

24) Sketch the graph of the equation $16x^2 + 25y^2 = 336$. Indicate the center, foci, and lengths of axes.

Answer: center $(2, 0)$, foci $(-1, 0)$ and $(5, 0)$, major axis 10, minor axis 8

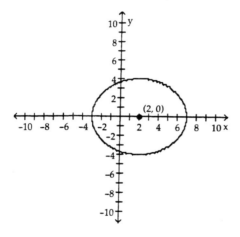

25) Sketch the graph of the equation $4x^2 + y^2 - 6y - 7 = 0$. Indicate the center, foci, and lengths of the axes.

Answer: center (0, 3), foci (0, 3 ± $\sqrt{5}$), major axis 8, minor axis 4

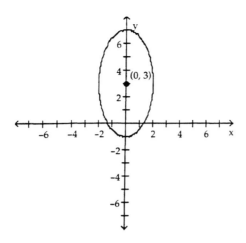

26) Sketch the graph of the equation $2x^2 + 3y^2 + 4x - 30y + 1 = 0$. Indicate the center, foci, and lengths of the axes.

Answer: center (−1, 5), foci (−1 ± $\sqrt{13}$, 5), major axis $2\sqrt{39}$, minor axis $2\sqrt{26}$

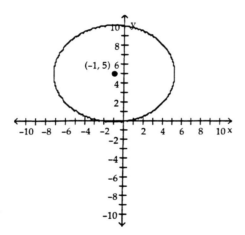

27) Find an equation for the hyperbola with foci (±5, 0) and vertices (±2, 0).

Answer: $\dfrac{x^2}{4} - \dfrac{y^2}{21} = 1$

28) Find an equation for the hyperbola with foci (±13, 0) and asymptotes $y = \pm \dfrac{5}{12}x$.

Answer: $\dfrac{x^2}{144} - \dfrac{y^2}{25} = 1$

29) Find an equation for the hyperbola with vertices (0, ±6) and asymptotes $y = \pm x$.

Answer: $\dfrac{y^2}{36} - \dfrac{x^2}{36} = 1$

30) Find an equation for the hyperbola with vertices (± 5, 0) and eccentricity $e = \frac{13}{5}$.

Answer: $\frac{x^2}{25} - \frac{y^2}{144} = 1$

31) Find an equation for the hyperbola with foci (0, ± 9) and eccentricity $e = 3$.

Answer: $\frac{y^2}{9} - \frac{x^2}{72} = 1$

32) Find an equation for the hyperbola with vertices (± 3, 0) and and passing through (12, 30).

Answer: $\frac{x^2}{9} - \frac{y^2}{60} = 1$

33) Find an equation for the hyperbola with foci (0, ± 4) and directrices $y = \pm \frac{1}{4}$.

Answer: $\frac{y^2}{1} - \frac{x^2}{15} = 1$

34) Find an equation for the hyperbola with center (3, 4), horizontal transverse axis of length 8, and eccentricity $e = 3$.

Answer: $\frac{(x-3)^2}{16} - \frac{(y-4)^2}{128} = 1$

35) Find an equation for the hyperbola with center (–2, –4), vertices (–6, –4) and (2, –4), and foci (–7, –4) and (3, 4).

Answer: $\frac{(x+2)^2}{16} - \frac{(y+4)^2}{9} = 1$

36) Find an equation for the hyperbola with center (–2, 1), vertices (–2, 3) and (–2, –1), and asymptotes $2x - 3y = -7$ and $2x + 3y = -1$.

Answer: $\frac{(x-1)^2}{4} - \frac{(y+2)^2}{9} = 1$

37) Find an equation for the hyperbola with focus (11, 2), and asymptotes $4x - 3y = 18$ and $4x + 3y = 30$.

Answer: $\frac{(x-6)^2}{9} - \frac{(y-2)^2}{16} = 1$

38) Sketch the graph of the hyperbola with the equation $x^2 - 3y^2 + 6x = 0$. Indicate the center, foci, and asymptotes.

Answer: center $(-3, 0)$; foci $(-3 \pm 2\sqrt{3}, 0)$; asymptotes $\sqrt{3}x - 3y = -3\sqrt{3}$ and $\sqrt{3}x + 3y = -3\sqrt{3}$

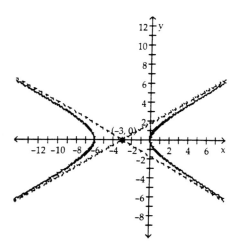

39) Sketch the graph of the hyperbola with the equation $y^2 - 2x^2 + 4y = 0$. Indicate the center, foci, and asymptotes.

Answer: center $(0, -2)$; foci $(0, -2 \pm \sqrt{6})$; asymptotes $\sqrt{2}x - y = 2$ and $\sqrt{2}x + y = -2$

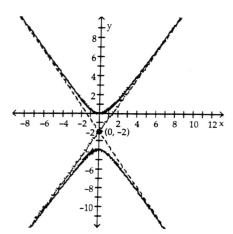

40) Sketch the graph of the hyperbola with the equation $16x^2 - 32x - 25y^2 - 50y = 409$. Indicate the center, foci, and asymptotes.

Answer: center $(1, -1)$; foci $(1 \pm \sqrt{41}, -1)$; asymptotes $4x - 5y = 9$ and $4x + 5y = 1$

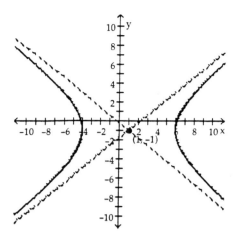

41) Sketch the graph of the hyperbola with the equation $25y^2 + 100y - 144x^2 - 864x = 4796$. Indicate the center, foci, and asymptotes.

Answer: center $(-3, -2)$; foci $(-3, 11)$ and $(-3, -15)$; asymptotes $12x - 5y = -26$ and $12x + 5y = -46$

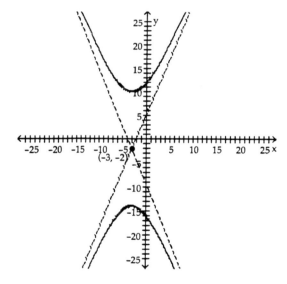

42) Identify and sketch the conic section with the polar equation $r = \dfrac{4}{1 + \cos \theta}$.

Answer: parabola

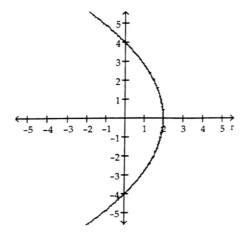

43) Identify and sketch the conic section with the polar equation $r = \dfrac{8}{4 + 2\cos \theta}$.

Answer: ellipse

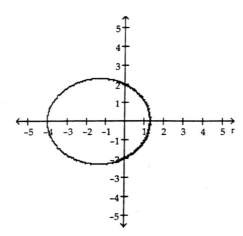

44) Identify and sketch the conic section with the polar equation $r = \dfrac{2}{2 - 3\cos\theta}$.

Answer: hyperbola

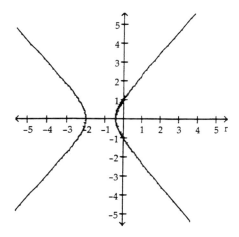

45) Find an equation of the parabola that has a vertical axis and passes through the points (–6, 2), (–4, 2), and (–1, –13).

Answer: $(x + 5)^2 = -(y - 3)$

46) A projectile is fired with initial velocity $v_0 = 30$ m/s from the origin at an angle of inclination $\alpha = 45°$. Use $g = 9.8$ m/s^2 to find the range and maximum height it attains. Round to two decimal places, if necessary.

Answer: R = 91.84 m; M = 22.96 m

47) A projectile is fired with initial velocity $v_0 = 50$ m/s from the origin at an angle of inclination α. Use $g = 9.8$ m/s^2 to find the value or values of α for which the range is 150 m.

Answer: $\alpha \approx \dfrac{\pi}{10}$

48) A projectile is fired with initial velocity $v_0 = 40$ m/s from the origin at an angle of inclination $\alpha = 30°$. Use $g = 9.8$ m/s^2 to find the length of time it remains above ground. Round to two decimal places, if necessary.

Answer: t = 4.09 s

49) The orbit of the comet Zoolan is an ellipse of extreme eccentricity e = 0.999945. The star Nebrackii is at one focus of this ellipse. The minimum distance Zoolan and Nebrackii is 0.12 AU. What is the maximum distance between Zoolan and Nebrackii? Round to two decimal places, if necessary.

Answer: 4363.52 AU

50) Find an equation of the ellipse with horizontal and vertical axes that pass through the points (–4, 0), (4, 0), (0, 1), and (0, –5).

Answer: Any equation for an ellipse centered at the origin, $\dfrac{x^2}{a^2} + \dfrac{y^2}{b^2} = 1$, with the restriction that $a \geq 4$ and $b \geq 5$, satisfies these conditions.

Ch. 11 Infinite Series

11.1 Introduction

1) There are no exercises for this section.

   Answer:

11.2 Infinite Sequences

Solve the problem.

1) Find a pattern in the sequence with the terms 1, 8, 27, 64, ... and (assuming that it continues as indicated) write a formula for the general term a_n of the sequence.

   Answer: $a_n = n^3$

2) Find a pattern in the sequence with the terms $\frac{1}{5}, \frac{1}{25}, \frac{1}{125}, \frac{1}{625}, \ldots$ and (assuming that it continues as indicated) write a formula for the general term a_n of the sequence.

   Answer: $a_n = \frac{1}{5^n}$

3) Find a pattern in the sequence with the terms $\frac{1}{3}, \frac{1}{5}, \frac{1}{7}, \frac{1}{9}, \ldots$ and (assuming that it continues as indicated) write a formula for the general term a_n of the sequence.

   Answer: $a_n = \frac{1}{2n+1}$

4) Find a pattern in the sequence with the terms 3, 1, 3, 1, ... and (assuming that it continues as indicated) write a formula for the general term a_n of the sequence.

   Answer: $a_n = 2 - (-1)^n$

5) Determine whether or not the sequence $a_n = \frac{3n}{7n-4}$ converges and find its limit if it does converge.

   Answer: $\frac{3}{7}$

6) Determine whether or not the sequence $a_n = 3 + (\frac{5}{6})^n$ converges and find its limit if it does converge.

   Answer: 3

7) Determine whether or not the sequence $a_n = 1 - (-2)^n$ converges and find its limit if it does converge.

   Answer: does not converge

8) Determine whether or not the sequence $a_n = \frac{\cos^2 n}{\sqrt[3]{n}}$ converges and find its limit if it does converge.

   Answer: 0

9) Determine whether or not the sequence $a_n = \frac{\ln 2n}{2\sqrt{n}}$ converges and find its limit if it does converge.

   Answer: 0

10) Determine whether or not the sequence $a_n = \left(\dfrac{n-1}{n+5}\right)^n$ converges and find its limit if it does converge.

   Answer: e^{-6}

11) Determine whether or not the sequence $a_n = \sqrt[n]{2^{2n+3}}$ converges and find its limit if it does converge.

   Answer: 4

12) Determine whether or not the sequence $a_n = \left(\dfrac{n}{3}\right)^{2/n}$ converges and find its limit if it does converge.

   Answer: 1

13) Determine whether or not the sequence $a_n = \left(\dfrac{5-n^2}{n^2-6}\right)^n$ converges and find its limit if it does converge.

   Answer: does not converge

14) Investigate the sequence $a_n = \sqrt{\dfrac{9n^2+2}{n^2+4n}}$ numerically or graphically. Formulate a reasonable guess for the value of its limit. Then apply limit laws to verify that your guess is correct.

   Answer: 3

15) Investigate the sequence $a_n = 2\tan^{-1}n^2$ numerically or graphically. Formulate a reasonable guess for the value of its limit. Then apply limit laws to verify that your guess is correct.

   Answer: π

11.3 Infinite Series and Convergence
Solve the problem.

1) One pane of a certain kind of glass transmits half the incident light, absorbs a quarter, and reflects a quarter. Two such panes are used in a double-pane window. What fraction of the incoming light will be transmitted?

   Answer: $\dfrac{4}{15}$

2) Consider whether the infinite series $1 + \dfrac{1}{2} + \dfrac{1}{4} + \dfrac{1}{2^n} + \ldots$ converges or diverges. If it converges, find its sum.

   Answer: 2

3) Consider whether the infinite series $1 - 3 + 9 - 27 + 81 \ldots (-3^n) + \ldots$ converges or diverges. If it converges, find its sum.

   Answer: diverges

4) Consider whether the infinite series $\sum\limits_{n=0}^{\infty} (-1)^n \left(\dfrac{7}{2e}\right)^n$ converges or diverges. If it converges, find its sum.

   Answer: diverges

5) Consider whether the infinite series $\sum_{n=1}^{\infty}\left(\dfrac{1}{2n}-\dfrac{1}{3n}\right)$ converges or diverges. If it converges, find its sum.

   Answer: diverges

6) Consider whether the infinite series $\sum_{n=0}^{\infty}\left(\dfrac{81}{80}\right)^n$ converges or diverges. If it converges, find its sum.

   Answer: diverges

7) Consider whether the infinite series $\sum_{n=0}^{\infty}\dfrac{1+3^n+4^n}{2^n}$ converges or diverges. If it converges, find its sum.

   Answer: diverges

8) Consider whether the infinite series $\sum_{n=0}^{\infty}\cos^n 2$ converges or diverges. If it converges, find its sum.

   Answer: $\dfrac{1}{1-\cos 2}$

9) Consider whether the infinite series $\sum_{n=1}^{\infty}(\arccos(-1))^n$ converges or diverges. If it converges, find its sum.

   Answer: diverges

10) Find the rational number represented by the repeating decimal 0.34343434...

    Answer: $\dfrac{34}{99}$

11) Find the rational number represented by the repeating decimal 2.718371837183...

    Answer: $\dfrac{2471}{909}$

12) Find the set of all those values of x for which the series $\sum_{n=1}^{\infty}\left(\dfrac{x}{3}\right)^{2n}$ is a convergent geometric series, then express the sum of the series as a function of x.

    Answer: $-3 < x < 3$; $\dfrac{x^2}{9-x^2}$

13) Find the set of all those values of x for which the series $\sum_{n=1}^{\infty}\left(\dfrac{x+1}{2}\right)^n$ is a convergent geometric series, then express the sum of the series as a function of x.

    Answer: $-\dfrac{x+1}{x-1}$; $-3 < x < 1$

14) Find the set of all those values of x for which the series $\sum_{n=1}^{\infty} \left(\frac{\cos x}{2}\right)^n$ is a convergent geometric series, then express the sum of the series as a function of x.

Answer: $\dfrac{\cos x}{2 - \cos x}$

15) Express the nth partial sum of the infinite series $\sum_{n=3}^{\infty} \dfrac{1}{n^2 - 4}$ as a telescoping sum and thereby find the sum of the series if it converges.

Answer: $S_n = -\dfrac{1}{4}\left(\dfrac{1}{n+1} + \dfrac{1}{n+2} + \dfrac{1}{n+3} + \dfrac{1}{n+4}\right) + \dfrac{25}{48}$; $\dfrac{25}{48}$

16) Express the nth partial sum of the infinite series $\sum_{n=3}^{\infty} \ln\left(\dfrac{n}{n+1}\right)$ as a telescoping sum and thereby find the sum of the series if it converges.

Answer: $S_n = -\ln(n+3) + \ln(3)$; diverges to $-\infty$

17) Express the nth partial sum of the infinite series $\sum_{n=2}^{\infty} \dfrac{1}{n(n-1)}$ as a telescoping sum and thereby find the sum of the series if it converges.

Answer: $S_n = -\dfrac{1}{n+1} + 1$; 1

18) A ball has bounce coefficient r<1 if, when it is dropped from a height h, it bounces back to a height of rh. A ball with a bounce coefficient 0.81 is dropped from an initial height of a = 4 ft. Use a geometric series to compute the total time required for it to complete its infinitely many bounces. The time required for a ball to drop h feet (from rest) is $\sqrt{2h/g}$ seconds, where g = 32 ft/s^2.

Answer: 9.5 seconds

19) A person sweeps up 1 pound of dirt from a walkway onto a shovel and drops 80% of it while carrying it over to a dumpster. She sweeps up what fell, again onto a shovel, and drops 80% of that on the way to the dumpster. This pattern continues. How many pounds of dirt will have been swept onto the shovel assuming an infinite number of sweepings?

Answer: 5 pounds

20) Moe, Larry, and Curley toss two fair coins in turn until one of them gets two heads. In other words, Moe tosses the two coins; if he doesn't win, Larry tosses the coins; and so on, until someone wins. Calculate the probability for each person to win.

Answer: Moe, $\dfrac{16}{37}$; Larry, $\dfrac{12}{37}$; Curley, $\dfrac{9}{37}$

11.4 Taylor Series and Taylor Polynomials
Solve the problem.

1) Write the Taylor series with center zero for the function $f(x) = \ln(1 + x^2)$.

   Answer: $\sum_{n=0}^{\infty} \dfrac{(-1)^n x^{2n+2}}{n+1}$

2) Write the Taylor series with center zero for the function $f(x) = \sqrt{x+1}$.

   Answer: $1 + \dfrac{x}{2} + \sum_{n=2}^{\infty} (-1)^{n-1} \dfrac{1 \cdot 3 \cdot 5 \cdot \ldots \cdot (2n-3)}{2^n n!} x^n$

3) Derive and write the Taylor series with center zero for the function $f(x) = e^{-x}$.

   Answer: $\sum_{n=0}^{\infty} \dfrac{(-1)^n x^n}{n!}$

4) Derive and write the Taylor series with center zero for the function $f(x) = \sin x$.

   Answer: $\sum_{n=0}^{\infty} \dfrac{(-1)^n x^{2n+1}}{(2n+1)!}$

5) Write the Taylor series with center 1 for the function $f(x) = x^4$.

   Answer: $1 + 4(x-1) + 6(x-1)^2 + 4(x-1)^3 + (x-1)^4$

6) Write the Taylor series with center zero for the function $f(x) = \ln\left(\dfrac{1+x}{1-x}\right)$.

   Answer: $\sum_{n=0}^{\infty} \dfrac{2x^{2n+1}}{2n+1}$

7) Write the Taylor series with center 2 for $g(x) = x^3$.

   Answer: $8 + 12(x-2) + 6(x-2)^2 + (x-2)^3$

8) Write the Maclaurin series for $h(x) = \sin x \cos x$.

   Answer: $\sum_{n=0}^{\infty} \dfrac{(-1)^n 4^n x^{2n+1}}{(2n+1)!}$

9) Write the Maclaurin series for $f(x) = \arctan x$.

   Answer: $\sum_{n=0}^{\infty} \dfrac{(-1)^n x^{2n+1}}{2n+1}$

10) Write the Taylor polynomial with center zero and of degree 5 for the function $f(x) = \sin x$.

    Answer: $x - \dfrac{1}{6}x^3 + \dfrac{1}{120}x^5$

11) Write the Taylor polynomial with center zero and of degree 4 for the function
$f(x) = e^{-x}$.

Answer: $1 - x + \dfrac{x^2}{2} - \dfrac{x^3}{6} + \dfrac{x^4}{24}$

12) Find Taylor's formula for the function f at a = 0. Find both the Taylor polynomial $P_n(x)$ of the indicated degree n and the remainder term $R_n(x)$.

$f(x) = \dfrac{1}{x+2}, n = 3$

Answer: $P_3(x) = \dfrac{1}{2} - \dfrac{1}{4}x + \dfrac{1}{8}x^2 - \dfrac{1}{16}x^3 + \dfrac{24}{(z+2)^5} \cdot \dfrac{x^4}{4!}$ for some z between 0 and x.

13) Find Taylor's formula for the function f at a = 0. Find both the Taylor polynomial $P_n(x)$ of the indicated degree n and the remainder term $R_n(x)$.

$f(x) = x^3 + 3x + 5, n = 4$

Answer: $P_4(x) = x^3 + 3x + 5$, $R_4(x) = 0$

14) Find the Taylor polynomial with remainder by using the given values of a and x.

$f(x) = e^{-x}$; $a = 1, n = 4$

Answer: $e^{-x} = e^{-1} + (-e^{-1})(x-1) + (\dfrac{1}{2}e^{-1})(x-1)^2 + (-\dfrac{1}{6}e^{-1})(x-1)^3 + (\dfrac{1}{24}e^{-1})(x-1)^4$

$+ \dfrac{-e^{-z}}{5!}(x-1)^5$ for some z between 1 and x.

15) Find the Taylor polynomial with remainder by using the given values of a and x.

$f(x) = \cos x$, $a = \dfrac{\pi}{2}, n = 4$

Answer: $\cos x = -(x - \dfrac{1}{2}\pi) + \dfrac{1}{6}(x - \dfrac{1}{2}\pi)^3 + \dfrac{-\sin z}{5!}(x - \dfrac{1}{2}\pi)^5$ for some z between $\dfrac{\pi}{2}$ and x.

16) Find the Taylor polynomial with remainder by using the given values of a and x.

$f(x) = \dfrac{1}{2x-1}$, $a = -3, n = 5$

Answer: $\dfrac{1}{2x-1} = -\dfrac{1}{7} - \dfrac{2}{49}(x+3) - \dfrac{4}{343}(x+3)^2 - \dfrac{8}{2401}(x+3)^3 - \dfrac{16}{16{,}807}(x+3)^4$

$- \dfrac{32}{117{,}649}(x+3)^5 + \dfrac{46{,}080(2z-1)^{-7}}{6!}(x+3)^6$ for some z between -3 and x.

17) Find the Taylor polynomial with remainder by using the given values of a and x.

$$f(x) = \frac{1}{\sqrt[3]{2x+1}} \quad a = 0, n = 3$$

Answer: $\dfrac{1}{\sqrt[3]{2x+1}} = 1 - \dfrac{2}{3}x + \dfrac{8}{9}x^2 - \dfrac{112}{81}x^3 + \dfrac{4480\,(\sqrt[3]{2z+1})^{-13}}{81 \cdot 4!}x^4$ for some z between 0 and x.

18) Find the Maclaurin series of the function f by substituting in the known series for e^x.

$f(x) = e^{x/2}$

Answer: $e^{x/2} = \sum\limits_{n=0}^{\infty} \dfrac{x^n}{n!\,2^n}$

19) Find the Maclaurin series of the function f by substituting in the known series for e^x.

$f(x) = e^{x^2/2}$

Answer: $e^{x^2/2} = \sum\limits_{n=0}^{\infty} \dfrac{x^{2n}}{n!\,2^n}$

20) Find the Maclaurin series of the function f by substituting in the known series for cos x.

$f(x) = \cos x^3$

Answer: $\cos x^3 = \sum\limits_{n=0}^{\infty} (-1)^n \dfrac{x^{6n}}{(2n)!}$

21) Find the Taylor series of the function at the indicated point a.

$f(x) = \cos x, \ a = \dfrac{\pi}{2}$

Answer: $\cos x = \sum\limits_{n=0}^{\infty} (-1)^{n+1} \dfrac{(x - \frac{1}{2}\pi)^{2n+1}}{(2n+1)!}$

22) Find the Taylor series of the function at the indicated point a.

$f(x) = e^{-x}, \ a = 1$

Answer: $e^{-x} = \sum\limits_{n=0}^{\infty} (-1)^n e^{-1} \dfrac{(x-1)^n}{n!}$

23) Find the Taylor series of the function at the indicated point a.

$f(x) = \ln x$, $a = 1$

Answer: $\ln x = \sum_{n=1}^{\infty} (-1)^{n+1} \frac{(x-1)^n}{n}$

11.5 The Integral Test
Solve the problem.

1) Test $\sum_{n=1}^{\infty} \frac{5n^2 + 1}{2n^3 - 1}$ for convergence.

   Answer: diverges

2) Test $\sum_{n=2}^{\infty} \frac{\ln n}{n}$ for convergence.

   Answer: diverges

3) Find the sum of the series $\sum_{n=1}^{\infty} \frac{1}{n(n+1)}$.

   Answer: $\ln 2$

4) Test for convergence. State what convergence test you use.

   (a) $\sum_{n=1}^{\infty} \frac{2^n}{n!}$ (b) $\sum_{n=1}^{\infty} 2^{-1/n}$ (c) $\sum_{n=1}^{\infty} \frac{10}{(n+1)^{3/2}}$

   Answer: (a) converges, ratio test (b) diverges, test for divergence
   (c) converges, comparison test with a p-series

5) Use the integral test to test the series $\sum_{n=1}^{\infty} \sin \frac{1}{n}$ for convergence.

   Answer: diverges

6) Use the integral test to test the series $\sum_{n=1}^{\infty} \left(\frac{1}{n} - \frac{1}{n^2} \right)$ for convergence.

   Answer: diverges

7) Use the integral test to test the series $\sum_{n=1}^{\infty} \frac{5+n}{n+1}$ for convergence.

   Answer: diverges

Calculus 247

8) Use the integral test to test the series $\sum_{n=1}^{\infty} \frac{\ln n}{n}$ for convergence.

   Answer: diverges

9) Use the integral test to test the series $\sum_{n=1}^{\infty} \frac{1}{n(\ln n)^{1.1}}$ for convergence.

   Answer: converges

10) Tell why the integral test does not apply to the series $\sum_{n=1}^{\infty} \frac{(-1)^{n+1}}{n^2}$.

    Answer: The series has both positive and negative terms.

11) Tell why the integral test does not apply to the series $\sum_{n=1}^{\pi} \frac{1 + n \cos n}{n}$.

    Answer: The series is not decreasing. Also, the series has positive and negative terms.

12) Determine the values of p for which the series $\sum_{n=1}^{\infty} \frac{1}{(2p)^n}$ converges.

    Answer: $p > \frac{1}{2}$, $p < -\frac{1}{2}$

13) Determine the values of p for which the series $\sum_{n=1}^{\infty} n e^{-pn}$ converges.

    Answer: $p > 0$

14) For the series $\sum_{n=1}^{\infty} \frac{1}{n^3}$ find the least positive integer n such that the remainder R_n in Theorem 2: The Integral Test Remainder Estimate is less than E = 0.00005.

    Answer: $n > 100$

15) For the series $\sum_{n=1}^{\infty} \frac{1}{n^5}$ find the least positive integer n such that the remainder R_n in Theorem 2: The Integral Test Remainder Estimate is less than E = 0.0000000025.

    Answer: $n > 100$

16) Find the sum of the series accurate to k decimal places. Begin by finding the smallest value of n such that the remainder satisfies the inequality $R_n < 5 \times 10^{-(k+1)}$. Then use a calculator to compute the partial sum S_n and round off appropriately.

$$\sum_{n=1}^{\infty} \frac{1}{n^8}, k = 8$$

Answer: 1.00407734

11.6 Comparison Tests for Positive-Term Series
Solve the problem.

1) Use a comparison test to determine whether the infinite series $\sum_{n=1}^{\infty} \frac{1}{n^3 + n - 1}$ converges or diverges.

Answer: converges

2) Use a comparison test to determine whether the infinite series $\sum_{n=1}^{\infty} \frac{1}{\sqrt{n} + 1}$ converges or diverges.

Answer: diverges

3) Use a comparison test to determine whether the infinite series $\sum_{n=1}^{\infty} \frac{1}{2 + e^n}$ converges or diverges.

Answer: converges

4) Use a comparison test to determine whether the infinite series $\sum_{n=1}^{\infty} \frac{n^2}{n + n^2 + n^{5/2}}$ converges or diverges.

Answer: diverges

5) Use a comparison test to determine whether the infinite series $\sum_{n=1}^{\infty} \sin \frac{1}{n^2}$ converges or diverges.

Answer: converges

6) Use a comparison test to determine whether the infinite series $\sum_{n=1}^{\infty} \frac{\cos^2 2n}{n^{3/2} + 1}$ converges or diverges.

Answer: converges

7) Use a comparison test to determine whether the infinite series $\sum_{n=1}^{\infty} \dfrac{1}{n^2 + \sin(1/n)}$ converges or diverges.

Answer: converges

8) Use a comparison test to determine whether the infinite series $\sum_{n=1}^{\infty} \dfrac{\arccos \frac{1}{n}}{n}$ converges or diverges.

Answer: diverges

9) Use a comparison test to determine whether the infinite series $\sum_{n=1}^{\infty} \dfrac{\left(1 + \frac{1}{n}\right)^n}{n}$ converges or diverges.

Answer: diverges

10) Use a comparison test to determine whether the infinite series $\sum_{n=1}^{\infty} \dfrac{n^2 + e^{-n} + 3}{n^3 + 5 + e^n}$ converges or diverges.

Answer: converges

11) Use a comparison test to determine whether the infinite series $\sum_{n=1}^{\infty} \dfrac{1}{\sqrt[3]{5n^5 + n + 2}}$ converges or diverges.

Answer: converges

12) Use a comparison test to determine whether the infinite series $\sum_{n=1}^{\infty} n^{-1} \ln\left(\dfrac{n+1}{n}\right)$ converges or diverges.

Answer: converges

13) Calculate the sum of the first ten terms of the series $\sum_{n=1}^{\infty} \dfrac{1}{n^3 + 3}$, then estimate the error made in using this partial sum to approximate the sum of the series.
Answer: 0.40874; 0.005

14) Calculate the sum of the first ten terms of the series $\sum_{n=1}^{\infty} \dfrac{\sin^2 n}{n^3}$, then estimate the error made in using this partial sum to approximate the sum of the series.
Answer: 0.83253; 0.005

15) Determine the smallest possible integer n such that the remainder satisfies the inequality $R_n < 0.005$. Then use a calculator or computer to approximate the sum of the series accurate to two decimal places.

$$\sum_{n=1}^{\infty} \frac{\cos \frac{1}{n}}{3^n}$$

Answer: 5; 0.33

16) Determine the smallest possible integer n such that the remainder satisfies the inequality $R_n < 0.005$. Then use a calculator or computer to approximate the sum of the series accurate to two decimal places.

$$\sum_{n=1}^{\infty} \frac{1}{n^{1+\sqrt{n}}}$$

Answer: 200; 1.26

11.7 Alternating Series and Absolute Convergence
Solve the problem.

1) Test $\sum_{n=1}^{\infty} \frac{n^3}{2^n}$ for convergence.

Answer: converges

2) Test $\sum_{n=1}^{\infty} \frac{(-1)^n \pi^{2n}}{(2n)!}$ for convergence.

Answer: converges

3) Test for convergence:

(a) $\sum_{n=1}^{\infty} \frac{n}{10n - 1}$ (b) $\sum_{n=1}^{\infty} \frac{2n}{n^4 + 1}$

(c) $\sum_{n=2}^{\infty} \frac{\ln n}{n^2}$ (d) $\sum_{n=2}^{\infty} \frac{n}{2n^2 + (\ln n)^2}$

(e) $\sum_{n=1}^{\infty} (n^{1/n} - 1)$

Answer: (a) diverges (b) converges (c) converges (d) diverges (e) diverges

4) Use the integral test to test the series $\sum_{n=1}^{\infty} \frac{n}{e^{n^2}}$ for convergence.

Answer: converges

Calculus 251

5) Give an example of a series that converges but which does not converge absolutely. Prove that your example has both of the required properties.

Answer: $\sum_{n=1}^{\infty} \frac{(-1)^n}{n}$

The Alternating Series Test shows that this series converges. However, $\sum_{n=1}^{\infty} \left| \frac{(-1)^n}{n} \right|$ is the harmonic series, which we know to be divergent.

6) Determine whether the alternating series $\sum_{n=1}^{\infty} (-1)^{n+1} \sin \frac{1}{n}$ converges or diverges.

Answer: converges

7) Determine whether the alternating series $\sum_{n=1}^{\infty} (-1)^{n+1} \frac{n+2}{\sqrt{n^3+3}}$ converges or diverges.

Answer: converges

8) Determine whether the alternating series $\sum_{n=1}^{\infty} (-1)^n \frac{\sqrt{n}}{5n}$ converges or diverges.

Answer: converges

9) Determine whether the alternating series $\sum_{n=1}^{\infty} \left(-1 + \frac{1}{n} \right)^n$ converges or diverges.

Answer: diverges

10) Determine whether the alternating series $\sum_{n=1}^{\infty} \cos n\pi \frac{1}{n^2}$ converges or diverges.

Answer: converges

11) Determine whether the alternating series $\sum_{n=1}^{\infty} (-1)^{n+1} \frac{5n}{n!}$ converges or diverges.

Answer: converges

12) Determine whether the series $\sum_{n=1}^{\infty} (-1)^{n+1} \frac{1}{n!}$ converges absolutely, converges conditionally, or diverges.

Answer: converges absolutely

13) Determine whether the series $\sum_{n=1}^{\infty} e^{-n^2}$ converges absolutely, converges conditionally, or diverges.

Answer: converges absolutely

14) Determine whether the series $\sum_{n=1}^{\infty} \frac{n!}{(5n)!}$ converges absolutely, converges conditionally, or diverges.

Answer: converges absolutely

15) Determine whether the series $\sum_{n=1}^{\infty} (-1)^{n+1} \sqrt[n]{n^2}$ converges absolutely, converges conditionally, or diverges.

Answer: diverges

16) Determine whether the series $\sum_{n=1}^{\infty} (-1)^n \sqrt{n} \sin \frac{1}{n}$ converges absolutely, converges conditionally, or diverges.

Answer: converges conditionally

17) Determine whether the series $\sum_{n=1}^{\infty} (-1)^n \frac{1 \cdot 3 \cdot 5 \cdots (2n+1)}{2 \cdot 4 \cdot 6 \cdots (2n)}$ converges absolutely, converges conditionally, or diverges.

Answer: diverges

18) Determine whether the series $\sum_{n=1}^{\infty} \frac{2 \cdot 4 \cdot 6 \cdots (2n)}{2^n (n+2)!}$ converges absolutely, converges conditionally, or diverges.

Answer: converges absolutely

19) Sum the indicated number of initial terms of the alternating series. Then apply the alternating series remainder estimate to estimate the error in approximating the sum of the series with this partial sum. Finally, approximate the sum of the series, writing precisely the number of decimal places that thereby are guaranteed to be correct after rounding.

$$\sum_{n=1}^{\infty} (-1)^{n+1} \sin \frac{1}{n^2}, 5 \text{ terms}$$

Answer: 0.7

20) Sum the indicated number of initial terms of the alternating series. Then apply the alternating series remainder estimate to estimate the error in approximating the sum of the series with this partial sum. Finally, approximate the sum of the series, writing precisely the number of decimal places that thereby are guaranteed to be correct after rounding.

$$\sum_{n=0}^{\infty} (-1)^n \frac{1}{(2n)!}, 3 \text{ terms}$$

Answer: 0.54

21) Sum the indicated number of initial terms of the alternating series. Then apply the alternating series remainder estimate to estimate the error in approximating the sum of the series with this partial sum. Finally, approximate the sum of the series, writing precisely the number of decimal places that thereby are guaranteed to be correct after rounding.

$$\sum_{n=1}^{\infty} \frac{(-1)^{n+1}}{n}, \text{ 9 terms}$$

Answer: 0.7

22) Sum enough terms (tell how many) to approximate the sum of the series $\sum_{n=1}^{\infty} \frac{(-1)^{n+1}}{n^2}$, writing the sum rounded to four correct decimal places.

Answer: 99 terms; 0.8225

23) Sum enough terms (tell how many) to approximate the sum of the series $\sum_{0}^{\infty} (-1)^n \frac{(0.1)^{2n+1}}{(2n+1)!}$, writing the sum rounded to five correct decimal places.

Answer: 2 terms; 0.09983

24) Sum enough terms (tell how many) to approximate the sum of the series $\frac{1}{1.1} = \sum_{n=0}^{\infty} (-0.1)^n$, writing the sum rounded to six correct decimal places.

Answer: 6 terms; 0.909091

11.8 Power Series
Solve the problem.

1) Find the interval of convergence:

(a) $\sum_{n=1}^{\infty} \frac{(x-1)^n}{n^2 3^n}$ (b) $\sum_{n=1}^{\infty} \frac{x^n}{n^2}$ (c) $\sum_{n=1}^{\infty} \frac{2^n(x-1)^n}{n}$ (d) $\sum_{n=1}^{\infty} \frac{(x-2)^n}{n \cdot 3^n}$

Answer: (a) [-2, 4] (b) [-1, 1] (c) $\left[\frac{1}{2}, \frac{3}{2}\right)$ (d) [-1, 5)

2) Expand $f(x) = \frac{4}{4 + x^2}$ into a power series in powers of x (by any method). Find the interval of convergence of the series.

Answer: $\sum_{n=0}^{\infty} \frac{(-1)^n x^{2n}}{4^n}$; (-2, 2)

3) Find the interval of convergence of the power series $\sum_{n=1}^{\infty} n^2 x^n$.

Answer: (-1, 1)

4) Find the interval of convergence of the power series $\sum_{n=1}^{\infty} n^n x^n$.

Answer: $\{0\}$

5) Find the interval of convergence of the power series $\sum_{n=1}^{\infty} \dfrac{(-1)^n x^{2n}}{n!}$.

Answer: $(-\infty, \infty)$

6) Find the interval of convergence of the power series $\sum_{n=1}^{\infty} \dfrac{2^n x^{2n}}{n^3}$.

Answer: $\left[-\dfrac{1}{\sqrt{2}}, \dfrac{1}{\sqrt{2}}\right]$

7) Find the interval of convergence of the power series $\sum_{n=1}^{\infty} \dfrac{(-1)^n 2^n x^n}{n}$.

Answer: $\left(-\dfrac{1}{2}, \dfrac{1}{2}\right]$

8) Find the interval of convergence of the power series $\sum_{n=0}^{\infty} (2x+1)^n$.

Answer: $(-1, 0)$

9) Find the interval of convergence of the power series $\sum_{n=1}^{\infty} \dfrac{n!}{n^n} x^n$.

Answer: $[-e, e)$

10) Find the interval of convergence of the power series $\sum_{n=1}^{\infty} \dfrac{(-1)^n (x+1)^n}{n}$.

Answer: $(-2, 0]$

11) Find the interval of convergence of the power series $\sum_{n=1}^{\infty} \dfrac{(-1)^n 3^n}{n!} (2x+1)^n$.

Answer: $(-\infty, \infty)$

12) Find the interval of convergence of the power series $\sum_{n=1}^{\infty} x^{n!}$.

Answer: $(-1, 1)$

Calculus 255

13) Use power series established in the Power Series section of the Chapter 10: Infinite Series to find a power series representation of the function $f(x) = x \ln(1+x)$. Then determine the radius of convergence of the resulting series.

Answer: $\sum_{n=1}^{\infty} \dfrac{(-1)^{n+1} x^{n+1}}{n}$, $R = 1$

14) Use power series established in the Power Series section of the Chapter 10: Infinite Series to find a power series representation of the function $f(x) = x \cos 3x$. Then determine the radius of convergence of the resulting series.

Answer: $\sum_{n=0}^{\infty} (-1)^n \dfrac{3^{2n} x^{2n+1}}{(2n)!}$, $R = \infty$

15) Use power series established in the Power Series section of the Chapter 10: Infinite Series to find a power series representation of the function $f(x) = e^{-x^3}$. Then determine the radius of convergence of the resulting series.

Answer: $\sum_{n=0}^{\infty} (-1)^n \dfrac{x^{3n}}{n!}$, $R = \infty$

16) Use power series established in the Power Series section of the Chapter 10: Infinite Series to find a power series representation of the function $f(x) = \sqrt{2x+1}$. Then determine the radius of convergence of the resulting series.

Answer: $1 + \sum_{n=1}^{\infty} 2^n \dfrac{\frac{1}{2}\left(\frac{1}{2}-1\right)\left(\frac{1}{2}-2\right)\cdots\left(\frac{1}{2}-n+1\right)}{n!} x^n$, $R = \dfrac{1}{2}$

17) Use power series established in the Power Series section of the Chapter 10: Infinite Series to find a power series representation of the function $f(x) = \dfrac{\cos(x) - 1}{x^2}$. Then determine the radius of convergence of the resulting series.

Answer: $\sum_{n=1}^{\infty} (-1)^n \dfrac{x^{2n-2}}{(2n)!}$, $R = \infty$

Issue: For items 15–18, check the R notation against the chapter content and answers given for previous edition.

18) Find a power series representation for the function $f(x) = \int_0^x \ln(1+t^2)\, dt$ by using termwise integration.

Answer: $\sum_{n=1}^{\infty} (-1)^{n+1} \dfrac{x^{2n+1}}{n(2n+1)}$

19) Find a power series representation for the function $f(x) = \int_0^x \cos - t^2 \, dt$ by using termwise integration.

Answer: $\sum_{n=0}^{\infty} (-1)^n \frac{x^{4n+1}}{(4n+1)(2n)!}$

20) Find a power series representation for the function $f(x) = \int_0^x \frac{t}{1-t^3} \, dt$ by using termwise integration.

Answer: $\sum_{n=0}^{\infty} n \frac{x^{3n+2}}{3n+2}$

11.9 Power Series Computations
Solve the problem.

1) Use Taylor polynomials or Taylor series to estimate $\int_0^{0.1} \sin(x^2) \, dx$ with error not exceeding 0.001. [Suggestion: Use the alternating series error estimate.]

   Answer: $\approx .00033$

2) Use power series to approximate the value of $\int_0^{0.1} \frac{1}{x^5+1} \, dx$ accurate to 7 decimal places.

   Answer: ≈ 0.0999998

3) Use power series to approximate the value of cos(0.001) accurate to 7 decimal places.

   Answer: ≈ 0.9999995

4) Use infinite series methods to approximate the value of $\int_0^1 \frac{\sin x}{x} \, dx$. Your answer should be accurate to 6 decimal places and should include an error analysis.

   Answer: ≈ 0.946083; error $\leq .0000003$

5) Use an infinite series to approximate $\sqrt{50}$ accurate to three decimal places.

   Answer: $\sqrt{50} = 7.071$ using that $\sqrt{49+1} = 7\sqrt{1 + \frac{1}{49}}$ and the binomial expansion for $\sqrt{1+x}$.

6) Use an infinite series to approximate $e^{-0.01}$ accurate to three decimal places.

   Answer:
   $e^{-0.01} = .990$ using the expansion for e^{-x}.

7) Use an infinite series to approximate ln 1.2 accurate to three decimal places.

   Answer: ln 1.2 = 0.182 using the expansion for ln (1 + x)

8) Use an infinite series to approximate $\sin \frac{\pi}{17}$ accurate to three decimal places.

   Answer: $\sin \frac{\pi}{17} = 0.184$ using the expansion for sin x.

9) Use an infinite series to approximate $\cos \frac{\pi}{10}$ accurate to three decimal places.

   Answer: $\cos \frac{\pi}{10} = 0.951$ using the expansion for cos x.

10) Use power series to approximate the value of the integral $\int_0^1 e^{-x^2} dx$ accurate to four decimal places.

    Answer: 0.7468

11) Use power series to approximate the value of the integral $\int_0^{1/2} \frac{\ln(1+t) - t}{t} dt$ accurate to four decimal places.

    Answer: −0.0516

12) Use power series rather than l'Hôpital's rule to evaluate the limit $\lim_{x \to 0} \frac{e^{x^2} - 1}{x^2}$.

    Answer: 1

13) Use power series rather than l'Hopital's rule to evaluate the limit $\lim_{x \to 0} \frac{\sin(x) - x}{x(e^{x^2} - 1)}$.

    Answer: $-\frac{1}{6}$

14) Use power series rather than l'Hopital's rule to evaluate the limit $\lim_{x \to 0} \left(\frac{1}{x} - \frac{1}{\ln(1+x)} \right)$.

    Answer: $-\frac{1}{2}$

15) Calculate sin 85° accurate to four decimal places using Taylor's formula for an appropriate function centered at $x = \frac{\pi}{2}$.

    Answer: 0.9962

16) Calculate cos 40° accurate to four decimal places using Taylor's formula for an appropriate function centered $x = \frac{\pi}{4}$.

    Answer: 0.7660

17) Calculate sin 87° accurate to five decimal places using Taylor's formula for an appropriate function centered at $x = \frac{\pi}{2}$.

   Answer: 0.99863

18) Calculate cos 42° accurate to five decimal places using Taylor's formula for an appropriate function centered at $x = \frac{\pi}{4}$.

   Answer: 0.74314

19) Determine the number of decimal places of accuracy the formula yields for $|x| \leq 0.1$.

   $$\cos x \approx 1 - \frac{x^2}{2} + \frac{x^4}{4!}$$

   Answer: Eight decimal places

20) Determine the number of decimal places of accuracy the formula yields for $|x| \leq 0.1$.

   $$e^{-x} \approx 1 - x + \frac{x^2}{2} - \frac{x^3}{6} + \frac{x^4}{4!}$$

   Answer: Seven decimal places

21) Determine the number of decimal places of accuracy the formula yields for $|x| \leq 0.1$.

   $$\frac{1}{1+x} \approx 1 - x + x^2 - x^3 + x^4$$

   Answer: Five decimal places

22) Determine the number of decimal places of accuracy the formula yields for $|x| \leq 0.1$.

   $$\frac{1}{(1+x)^2} = 1 - 2x + 3x^2 - 4x^3 + 5x^4 - 6x^5$$

   Answer: Six decimal places

23) Approximate the area of the region that lies between the graph of $y = \frac{e^{x^2} - 1}{x^2}$ and the x-axis from $x = -1$ to $x = 1$. Then approximate the volume of the solid generated by rotating the region around the y-axis. Finally, approximate the volume of the solid generated by rotating the region around the x-axis.

   [Insert graph of $y = \frac{e^{x^2} - 1}{x^2}$ Graph style: x = [-1, 1], y = [0, 2]; blank background, labeled ticks at (0,0), y = [0.5, 1, 1.5, 2], x = [-1, -0.5, 0.5, 1]. Shade in area between plot and x-axis]

   Answer: area: 2.41
   volume around x-axis: 9.41
   volume around y-axis: 4.14

11.10 Series Solutions of Differential Equations
Solve the problem.

1) Find a power series solution of the differential equation $y' - 2y = 0$. Determine the radius of convergence of the resulting series, and use your knowledge of familiar Maclaurin series and the binomial series to identify the series solution in terms of familiar elementary functions.

   Answer: $y = c_0 e^{2x}$

2) Find a power series solution of the differential equation $3y' + 2y = 0$. Determine the radius of convergence of the resulting series, and use your knowledge of familiar Maclaurin series and the binomial series to identify the series solution in terms of familiar elementary functions.

   Answer: $c_0 e^{-2x/3}$

3) Find a power series solution of the differential equation $(x - 3)y' + y = 0$. Determine the radius of convergence of the resulting series, and use your knowledge of familiar Maclaurin series and the binomial series to identify the series solution in terms of familiar elementary functions.

   Answer: $\dfrac{c_0}{x - 3}$

4) Find a power series solution of the differential equation $(3x - 2)y' + 3y = 0$. Determine the radius of convergence of the resulting series, and use your knowledge of familiar Maclaurin series and the binomial series to identify the series solution in terms of familiar elementary functions.

   Answer: $\dfrac{c_0}{3x - 2}$

5) Find a power series solution of the differential equation $(x - 3)y' + 2y = 0$. Determine the radius of convergence of the resulting series, and use your knowledge of familiar Maclaurin series and the binomial series to identify the series solution in terms of familiar elementary functions.

   Answer: $c_0 \dfrac{1}{(x-3)^2}$

6) Use the method of Example 4 in Chapter 10: Series Solutions of Differential Equations to find two linearly independent power series solutions of the differential equation $y'' = 9y$. Determine the radius of convergence of each series, and identify the general solution in terms of familiar elementary functions.

   Answer: $y = c_0 \sum\limits_{n=0}^{\infty} 3^{2n} \dfrac{x^{2n}}{(2n)!} + c_1 \sum\limits_{n=0}^{\infty} 3^{2n+1} \dfrac{x^{2n+1}}{(2n+1)!} = c_0 \cosh 3x + c_1 \sinh 3x$

7) Use the method of Example 4 in Chapter 10: Series Solutions of Differential Equations to find two linearly independent power series solutions of the differential equation $y'' = -4y$. Determine the radius of convergence of each series, and identify the general solution in terms of familiar elementary functions.

   Answer: $y = c_0 \sum\limits_{n=0}^{\infty} (-1)^n 2^{2n} \dfrac{x^{2n}}{(2n)!} + c_1 \sum\limits_{n=0}^{\infty} (-1)^n 2^{(2n+1)} \dfrac{x^{2n+1}}{(2n+1)!} = c_0 \cos 2x + c_1 \sin 2x$

8) Show that the power series method fails to yield a power series solution of the form $y = \sum c_n x^n$ for the differential equation $x^2 y'' = -4y$.

   Answer: We do not get a recursion formula for c_n.

9) Show that the power series method fails to yield a power series solution of the form $y = \sum c_n x^n$ for the differential equation $x^3 y' + y = 0$.

   Answer: The series converges only for $x = 0$.

10) First derive a recurrence relation giving c_n for $n \geq 2$ in terms of c_0 or c_1 (or both). Then apply the initial conditions to find the values of c_0 and c_1. Next determine c_n in terms of n, as in the text) and, finally, identify the particular solution in terms of familiar elementary functions.

   $y'' + 9y = 0; \ y(0) = 2, \ y'(0) = 0$

   Answer: $y = 2 \sum_{n=0}^{\infty} (-1)^n \frac{x^{2n}}{(2n)!} = 2 \cos 3x$

11) First derive a recurrence relation giving c_n for $n \geq 2$ in terms of c_0 or c_1 (or both). Then apply the initial conditions to find the values of c_0 and c_1. Next determine c_n in terms of n, as in the text) and, finally, identify the particular solution in terms of familiar elementary functions.

   $y'' - 9y = 0; \ y(0) = 0, \ y'(0) = 3$

   Answer: $y = 3 \sum_{n=0}^{\infty} \frac{x^{2n+1}}{(2n+1)!} = \sinh 3x$

12) First derive a recurrence relation giving c_n for $n \geq 2$ in terms of c_0 or c_1 (or both). Then apply the initial conditions to find the values of c_0 and c_1. Next determine c_n in terms of n, as in the text) and, finally, identify the particular solution in terms of familiar elementary functions.

   $y'' - 4y' + 4y = 0; \ y(0) = 1, \ y'(0) = 2$

   Answer: $y = \sum_{n=0}^{\infty} \frac{2^n x^n}{n!} = e^{2x}$

13) First derive a recurrence relation giving c_n for $n \geq 2$ in terms of c_0 or c_1 (or both). Then apply the initial conditions to find the values of c_0 and c_1. Next determine c_n in terms of n, as in the text) and, finally, identify the particular solution in terms of familiar elementary functions.

   $y'' - 3y' + 2y = 0; \ y(0) = 1, \ y'(0) = 1$

   Answer: $y = \sum_{n=0}^{\infty} \frac{x^n}{n!} = e^x$

Ch. 12 Vectors, Curves, and Surfaces in Space
12.1 Vectors in the Plane
Solve the problem.

1) Find a vector $v = \langle a, b \rangle$ that is represented by $\overrightarrow{RS}$ where $R = (2, 5)$ and $S = (4, 1)$.

   Answer: $v = \langle 2, -4 \rangle$

2) Find a vector $v = \langle a, b \rangle$ that is represented by $\overrightarrow{RS}$ where $R(6, 12)$ and $S(-6, -12)$.

   Answer: $v = \langle -12, -24 \rangle$

3) Given $u = \langle 5, -3 \rangle$ and $v = \langle 4, 6 \rangle$, find $w = u + v$.

   Answer: $w = \langle 9, 3 \rangle$

4) Given $u = 6i + 9j$ and $v = 3i - 2j$, find $w = u + v$.

   Answer: $w = 9i + 7j$

5) Given $a = \langle 4, 5 \rangle$ and $b = \langle -5, 4 \rangle$, find (a) $|a|$, (b) $|-2b|$, (c) $|a - b|$, (d) $a + b$, and (e) $3a - 2b$.

   Answer: (a) $\sqrt{41}$; (b) $2\sqrt{41}$; (c) $\sqrt{82}$; (d) $\langle -1, 9 \rangle$; (e) $\langle 22, 7 \rangle$

6) Given $a = \langle -4, -4 \rangle$ and $b = \langle -5, -6 \rangle$, find (a) $|a|$, (b) $|-2b|$, (c) $|a - b|$, (d) $a + b$, and (e) $3a - 2b$.

   Answer: (a) $4\sqrt{2}$; (b) $2\sqrt{61}$; (c) $\sqrt{5}$; (d) $\langle -9, -10 \rangle$; (e) $\langle -2, 0 \rangle$

7) Given $a = 4i + 3j$ and $b = 3i - 2j$, find (a) $|a|$, (b) $|-2b|$, (c) $|a - b|$, (d) $a + b$, and (e) $3a - 2b$.

   Answer: (a) 5; (b) $2\sqrt{13}$; (c) $\sqrt{26}$; (d) $7i + j$; (e) $6i + 13j$

8) Given $a = 6i$ and $b = -5j$, find (a) $|a|$, (b) $|-2b|$, (c) $|a - b|$, (d) $a + b$, and (e) $3a - 2b$.

   Answer: (a) 6; (b) 10; (c) $\sqrt{61}$; (d) $6i - 5j$; (e) $18i + 10j$

9) Given $a = \langle -5, 12 \rangle$, find (a) a unit vector u with the same direction as a and (b) a unit vector v with the direction opposite that of a.

   Answer: (a) $u = \langle -\frac{5}{13}, \frac{12}{13} \rangle$; (b) $v = \langle \frac{5}{13}, -\frac{12}{13} \rangle$

10) Given $a = 6i + 9j$, find (a) a unit vector u with the same direction as a and (b) a unit vector v with the direction opposite that of a. (Express u and v in terms of i and j.)

    Answer: (a) $u = \frac{2\sqrt{117}}{39}i + \frac{\sqrt{117}}{13}j$; (b) $v = -\frac{2\sqrt{117}}{39}i - \frac{\sqrt{117}}{13}j$

11) Given $P = (6, 4)$ and $Q = (6, -5)$, find the vector a (in terms of i and j) that is represented by $\overrightarrow{PQ}$.

    Answer: $a = -9j$

12) Given $P = (-5, 10)$ and $Q = (5, -10)$, find the vector a (in terms of i and j) that is represented by $\overrightarrow{PQ}$.

    Answer: $a = 10i - 20j$

13) Given $a = \langle 5, 0 \rangle$ and $b = \langle 0, 9 \rangle$, determine whether or not a and b are perpendicular.

    Answer: yes, $a \perp b$

14) Given **a** = 5i + 9j and **b** = 12i – 10j, determine whether or not **a** and **b** are perpendicular.

   Answer: no, **a** is not ⊥ **b**

15) Given **a** = 4i + 5j and **b** = 6i + 7j, express i and j in terms of **a** and **b**.

   Answer: i = $\frac{5}{2}$**b** – $\frac{7}{2}$**a**; j = **b** – $\frac{6}{5}$**a**

16) Given **a** = 9i – 5j and **b** = 7i – 4j, express i and j in terms of **a** and **b**.

   Answer: i = 9**b** + 7**a**; j = 7**a** – 9**b**

17) Given **a** = i – j, **b** = i + j, and **c** = 3i + 2j express **c** in the form r**a** + s**b** where r and s are scalars.

   Answer: **c** = $\frac{1}{2}$**a** + $\frac{5}{2}$**b**

18) Given **a** = 2i + 3j, **b** = 3i + 5j, and **c** = 8i + 11j express **c** in the form r**a** + s**b** where r and s are scalars.

   Answer: **c** = 7**a** – 2**b**

19) Find a vector that has the same direction as 6i – 3j and is (a) four times its length; (b) one third its length.

   Answer: (a) 24i – 12j; (b) 2i – j

20) Find a vector that has opposite direction as 6i – 5j and is (a) three times its length; (b) one half its length.

   Answer: (a) –18i + 15j; (b) –3i + $\frac{5}{2}$j

21) Find a vector of length 3 with (a) the same direction as 5i – 4j; (b) the direction opposite that of –2i + 6j.

   Answer: (a) $\frac{15i\sqrt{41}}{41} - \frac{12j\sqrt{41}}{41}$; (b) $\frac{3i\sqrt{10}}{10} - \frac{9j\sqrt{10}}{10}$

22) For what values of c are the vectors <2c, 4> and <c, –4> perpendicular?

   Answer: c = ± 2$\sqrt{2}$

23) For what values of c are the vectors 3ci – 2j and 4i + cj perpendicular?

   Answer: c = 0

12.2 Three Dimensional Vectors
Solve the problem.

1) Given **a** = <1, 3, –5> and **b** = <–1, 2, –3>, find (a) 2**a** + **b**; (b) 3**a** – 4**b**; (c) **a**•**b**; (d) |**a** – **b**|; (e) $\frac{\mathbf{a}}{|\mathbf{a}|}$; (f) the measure of the angle θ between **a** and **b**; (g) comp$_\mathbf{a}$ **b**; (h) comp$_\mathbf{b}$ **a**.

   Answer: (a) <1, 8, –13>; (b) <7, 1, –3>; (c) 20; (d) $\sqrt{17}$; (e) $\left\langle \frac{\sqrt{35}}{35}, \frac{3\sqrt{35}}{35}, -\frac{\sqrt{35}}{7} \right\rangle$; (f) θ = 25° = $\frac{5\pi}{36}$; (g) $\frac{10\sqrt{14}}{7}$; (h) $\frac{4\sqrt{35}}{7}$

2) Given $\mathbf{a} = \mathbf{i} - \mathbf{j} + \mathbf{k}$ and $\mathbf{b} = \mathbf{j} - \mathbf{k}$, find (a) $2\mathbf{a} + \mathbf{b}$; (b) $3\mathbf{a} - 4\mathbf{b}$; (c) $\mathbf{a} \cdot \mathbf{b}$; (d) $|\mathbf{a} - \mathbf{b}|$; (e) $\dfrac{\mathbf{a}}{|\mathbf{a}|}$; (f) the measure of the angle θ between $\mathbf{a}$ and $\mathbf{b}$; (g) $\text{comp}_\mathbf{a} \mathbf{b}$; (h) $\text{comp}_\mathbf{b} \mathbf{a}$.

Answer: (a) $2\mathbf{i} - \mathbf{j} + \mathbf{k}$; (b) $3\mathbf{i} - 7\mathbf{j} + 7\mathbf{k}$; (c) -2; (d) 3; (e) $\left\langle \dfrac{\sqrt{3}}{3}, -\dfrac{\sqrt{3}}{3}, \dfrac{\sqrt{3}}{3} \right\rangle$; (f) $\theta = 145° = \dfrac{29\pi}{36}$; (g) $-\sqrt{2}$; (h) $-\dfrac{2\sqrt{3}}{3}$

3) Given $\mathbf{a} = 5\mathbf{i} - \mathbf{j}$ and $\mathbf{b} = 3\mathbf{j} - 2\mathbf{k}$, find (a) $2\mathbf{a} + \mathbf{b}$; (b) $3\mathbf{a} - 4\mathbf{b}$; (c) $\mathbf{a} \cdot \mathbf{b}$; (d) $|\mathbf{a} - \mathbf{b}|$; (e) $\dfrac{\mathbf{a}}{|\mathbf{a}|}$; (f) the measure of the angle θ between $\mathbf{a}$ and $\mathbf{b}$; (g) $\text{comp}_\mathbf{a} \mathbf{b}$; (h) $\text{comp}_\mathbf{b} \mathbf{a}$.

Answer: (a) $10\mathbf{i} + \mathbf{j} - 2\mathbf{k}$; (b) $15\mathbf{i} - 15\mathbf{j} + 8\mathbf{k}$; (c) -3; (d) $3\sqrt{5}$; (e) $\left\langle \dfrac{5\sqrt{26}}{26}, -\dfrac{\sqrt{26}}{26}, 0 \right\rangle$; (f) $\theta = 99° = \dfrac{11\pi}{20}$; (g) $-\dfrac{3\sqrt{13}}{13}$; (h) $\dfrac{3\sqrt{26}}{26}$

4) Find the equation of the sphere with center $(-1, 3, -6)$ and radius $\sqrt{5}$.
Answer: $x^2 + 2x + y^2 - 6y + z^2 + 12z + 41 = 0$

5) Find the equation of the sphere with center $(-1, 3, -6)$ and passing through the point $(1, 1, 1)$.
Answer: $x^2 - 12x + y^2 + 4y + z^2 - 6z + 11 = 0$

6) Find the center and radius of the sphere with the equation $x^2 - 12x + y^2 + 8y + z^2 - 2z + 43 = 0$
Answer: center $(6, -4, 1)$; radius $\sqrt{10}$

7) Find the center and radius of the sphere with the equation $x^2 + y^2 + z^2 = 6y + 18z - 2x$.
Answer: center $(-1, 3, 9)$; radius $\sqrt{91}$

8) Describe the graph of the equation $y = 5$ using plain, clear language.
Answer: The vertical plane (∥ to the xz plane) meeting the y-axis at the point $(0, 5, 0)$.

9) Describe the graph of the equation $x^2 + y^2 + z^2 = 4x - 2y + 8z - 21$ using plain, clear language.
Answer: The single point $(2, -1, 4)$.

10) Given $\mathbf{a} = \langle 4, 5, 3 \rangle$ and $\mathbf{b} = \langle 4, -2, 2 \rangle$, determine whether $\mathbf{a}$ and $\mathbf{b}$ are parallel, perpendicular, or neither.
Answer: Perpendicular since $\mathbf{a} \cdot \mathbf{b} = 0$

11) Given $\mathbf{a} = 24\mathbf{i} - 8\mathbf{j} + 12\mathbf{k}$ and $\mathbf{b} = 3\mathbf{i} - \mathbf{j} + \dfrac{3}{2}\mathbf{k}$, determine whether $\mathbf{a}$ and $\mathbf{b}$ are parallel, perpendicular, or neither.
Answer: Parallel since $\mathbf{a} = 8\mathbf{b}$

12) Do the points $P(0, -5, 1)$, $Q(1, 10, 4)$, and $R(6, 8, -1)$ lie in a straight line?
Answer: No

13) Find (to the nearest degree) each angle of the triangle with vertices $A(1, 1, 1)$, $B(0, 0, 3)$, and $C(4, -1, -3)$.
Answer: $133°, 30°, 17°$

14) Find (to the nearest degree) each angle of the triangle with vertices $A(1, 2, 3)$, $B(4, 1, 3)$, and $C(1, 2, 6)$.
Answer: $90°, 47°, 43°$

15) Given the points P(1, 5, 3) and Q(2, 4, 1), find the direction angle of the vector represented by $\overrightarrow{PQ}$.

Answer: $\alpha \approx 48°$; $\beta \approx 109°$; $\gamma \approx 131°$

16) Given the points P(-3, -4, -5) and Q(3, 4, 5), find the direction angle of the vector represented by $\overrightarrow{PQ}$.

Answer: $\alpha \approx 64°$; $\beta \approx 56°$; $\gamma = 45°$ exactly

17) Given **F** = 4**i** - 2**k**; P(0, 1, 0) and Q(4, 0, 1), find the work W done by the force **F** in moving a particle in a straight line from P to Q.

Answer: W = 14

18) Given **F** = 3**i** - 6**j** + 2**k**; P(9, 3, 2) and Q(2, 1, -4), find the work W done by the force **F** in moving a particle in a straight line from P to Q.

Answer: W = -21

19) A rectangular solid in the first octant has vertices P(0, 0, 0), Q(2, 0, 0), R(2, 1, 0), and S(2, 1, 3) where P and S are opposites. Find the angle α between the edge of the cube on the x-axis and the diagonal $\overrightarrow{PS}$.

Answer: $\alpha \approx 58°$

12.3 The Cross Product of Vectors
Solve the problem.

1) Given **a** = <4, -3, -1> and **b** = <1, 4, 6>, find **a** × **b**.

Answer: <-14, -25, 19>

2) Given **a** = 2**i** - **j** + **k** and **b** = **i** + 3**j** + 2**k**, find **a** × **b**.

Answer: -5**i** - 3**j** + 7**k**

3) Given **a** = <1, -4> and **b** = <5, 7>, find **a** × **b**. (Note: expand **a** and **b** into 3 dimensional vectors to find **a** × **b**)

Answer: <0, 0, 27>

4) Given **a** = 4**i** - 6**j** and **b** = -3**i** - 9**j**, find **a** × **b**. (Note: expand **a** and **b** into 3 dimensional vectors to find **a** × **b**)

Answer: -54**k**

5) Given **a** = <1, 2, 3> and **b** = <4, 5, 6>, find two different unit vectors, **u** and **v**, both of which are perpendicular to **a** and to **b**.

Answer: **u** = <$-\frac{\sqrt{6}}{6}, \frac{\sqrt{6}}{3}, -\frac{\sqrt{6}}{6}$>; **v** = <$\frac{\sqrt{6}}{6}, -\frac{\sqrt{6}}{3}, \frac{\sqrt{6}}{6}$>

6) Given **a** = 2**i** + 3**j** + 4**k** and **b** = 3**i** - 4**j** - 5**k**, find two different unit vectors, **u** and **v**, both of which are perpendicular to **a** and to **b**.

Answer: **u** = $\frac{\sqrt{86}}{258}$**i** + $\frac{11\sqrt{86}}{129}$**j** - $\frac{17\sqrt{86}}{258}$**k**; **v** = -$\frac{\sqrt{86}}{258}$**i** - $\frac{11\sqrt{86}}{129}$**j** + $\frac{17\sqrt{86}}{258}$**k**

7) Given the rectangular vertices P(1, 2, 3), Q(3, 4, 5), and R(0, 1, 0), find (a) the area of the triangle formed and (b) the volume of the parallelepiped with adjacent edges $\overrightarrow{OP}$, $\overrightarrow{OQ}$, and $\overrightarrow{OR}$. (Note: O(0, 0, 0))

Answer: (a) A = $2\sqrt{2}$; (b) V = 4

8) Given the rectangular vertices P(−1, 2, −3), Q(−2, 3, −4), and R(0, 0, 1), find (a) the area of the triangle formed; (b) the volume of the parallelepiped with adjacent edges $\overrightarrow{OP}$, $\overrightarrow{OQ}$, and $\overrightarrow{OR}$; (c) the volume of the pyramid with vertices O, P, Q, and R; (d) a unit vector **n** perpendicular to the plane through P, Q, and R; and (e) the distance from the origin to this plane by computing **n**·$\overrightarrow{OP}$.(Note: O(0, 0, 0))

Answer: (a) $A = \frac{\sqrt{14}}{2}$; (b) $V = 1$; (c) $V = \frac{1}{6}$; (d) $<\frac{\sqrt{14}}{7}, \frac{3\sqrt{14}}{7}, \frac{\sqrt{14}}{14}>$; (e) $\frac{\sqrt{14}}{14}$

9) Given the points A(3, 1, 2), B(4, 2, 1), C(5, 4, 0) and D(7, 6, 6), determine whether or not the points are coplanar. If not, find the volume of the pyramid formed with these points as its vertices.

Answer: coplanar

10) Given the points A(1, 3, −4), B(3, 8, 6), C(−1, 2, −4) and D(−4, 4, −5), determine whether or not the points are coplanar. If not, find the volume of the pyramid formed with these points as its vertices.

Answer: not coplanar; V = 13

12.4 Lines and Planes in Space
Solve the problem.

1) Write the parametric equations of the straight line that passes through P(3, 2, 1) and is parallel to **v = i + 2j + 3k**.

Answer: $x = 3 + t$; $y = 2 + 2t$; $z = 1 + 3t$

2) Write the parametric equations of the straight line that passes through $P_1(2, 4, 6)$ and $P_2(8, 1, 3)$.

Answer: $x = 2 + 6t$; $y = 4 − 3t$; $z = 6 − 3t$

3) Write the parametric and symmetric equations of the straight line that passes through $P_1(1, 5, −3)$ and is perpendicular to the plane with equation $3x + y − 2z = 6$.

Answer: $x = 1 + 3t$; $y = 5 + t$; $z = 3 − 2t$; $\frac{x-1}{3} = \frac{y-5}{1} = \frac{-z-3}{2}$

4) Given L_1: $x = 5 − t$, $y = −5 + t$, $z = −9 + 2t$ and L_2: $x = −3 + s$, $y = 3 − s$, $z = −17 + 4s$, determine whether the lines are parallel, intersecting or skew. If they intersect, find the point of intersection.

Answer: intersecting at (1, −1, 1)

5) Given L_1: $\frac{1}{2}(x − 6) = \frac{1}{3}(y − 12) = \frac{1}{7}(z − 4)$ and L_2: $\frac{1}{8}(x − 1) = \frac{1}{24}(y + 3) = \frac{1}{28}(z − 5)$, determine whether the lines are parallel, intersecting or skew. If they intersect, find the point of intersection.

Answer: parallel

6) Find an equation of the plane with normal vector **n = 3i − 2j + k** that passes through P(6, 2, −5).

Answer: $3x − 2y + z = 9$

7) Find an equation of the plane through the points P(1, 3, 5), Q(2, 4, 6), and R(−5, −2, −3).

Answer: $3x − 2y − z − 8 = 0$

8) Write an equation of the plane that contains the point P(1, 3, 5) and the line L: $x = 4t$, $y = 6 + 5t$, $z = 3 − 2t$.

Answer: $4x + 8y + 18z = 118$

9) Determine whether the line L: $x = 5 - 4t$, $y = 16 + 6t$, $z = 2 + 5t$ and the plane $4x + y + 2z = 10$ intersect or are parallel. If the intersect, find the point of intersection.

   Answer: they are parallel

10) Given the planes $x + 2y + z = 4$ and $2x - 3y - z = 1$, find the angle between the planes and the parametric and symmetric equations of the line of intersection of the planes.

    Answer: $\theta \approx 57°$; $x = 1 + t$, $y = -2 + 3t$, $z = 7 - 7t$; $\frac{x-1}{1} = \frac{y+2}{3} = \frac{7-z}{7}$

11) Find an equation of the plane that passes through the points P(1, 2, 3) and Q(-1, 2, -3) and is parallel to line of intersection of the planes $x + y + z = 1$ and $4x - 2y - z = 5$.

    Answer: $-30x + 18y + 10z = 36$

12) Find the shortest distance between line L_1: $x = 1 + 2t$, $y = 3 - 4t$, $z = 2 + t$ and L_2: the intersection of the planes $x + y + z = 1$ and $2x + y - 3z = 10$.

    Answer: ≈ 1.46588

12.5 Curves and Motion in Space
Solve the problem.

1) Find the values of $\mathbf{r}'(t)$ and $\mathbf{r}''(t)$ for $\mathbf{r} = 2\cos(2t)\mathbf{i} - \sin(2t)\mathbf{j}$ at the point $t = \frac{\pi}{3}$.

   Answer: $\mathbf{r}'(\frac{\pi}{3}) = -2\sqrt{3}\mathbf{i} + 2\mathbf{j}$; $\mathbf{r}''(\frac{\pi}{3}) = 4\mathbf{i} + 4\sqrt{3}\mathbf{j}$

2) Evaluate the integral $\int_0^{\sqrt{\pi}/2} t\sin t^2 \mathbf{i}\, dt$.

   Answer: $\left(\frac{2-\sqrt{2}}{4}\right)\mathbf{j}$

3) Compute the derivative $D_t[\mathbf{u}(t)\cdot\mathbf{v}(t)]$ for $\mathbf{u}(t) = t^2\mathbf{i} - 2\mathbf{j}$ and $\mathbf{v}(t) = 5\mathbf{i} + 3t\mathbf{j}$.

   Answer: $10t\mathbf{i} - 6\mathbf{j}$

4) The acceleration vector for a particle moving in xyz-space is $\mathbf{a}(t) = -3\mathbf{j} + 2t\mathbf{k}$. The initial velocity is $\mathbf{v}_0 = \mathbf{j}$ and the initial position is $\mathbf{r}_0 = 2\mathbf{i}$. Find the position vector $\mathbf{r}(t)$ at time t.

   Answer: $\mathbf{r}(t) = 2\mathbf{i} + (t - \frac{3t^2}{2})\mathbf{j} + \frac{t^3}{3}\mathbf{k}$

5) A projectile is fired from the origin with initial speed $v_0 = 120$ ft/sec. Find the maximum height, y_{max}, and the range, R, of the projectile if the initial angle of inclination is (a) $\alpha = 30°$ (b) $\alpha = 45°$ (c) $\alpha = 60°$.
   $x(t) = v_0\cos(\alpha)t + x_0$, $y(t) = -0.5gt^2 + v_0\sin(\alpha)t + y_0$

   Answer: (a) $R = 225\sqrt{3} \approx 389.7$ ft; max height = 56.25 ft
   (b) $R = 450$ ft; max height 112.5 ft
   (c) $r = 225\sqrt{3} \approx 389.7$ ft; max height = 168.75 ft

6) A baseball is thrown with an initial velocity of 120 ft/sec straight upward from the ground. It experiences a downward gravitational acceleration of 32 ft/sec². Because of spin it also experiences a (horizontal) northward acceleration of 0.2 ft/sec². Otherwise, the air has no effect on its motion. How far north of the throwing point will the ball land?

Answer: 5.625 ft

7) A particle moves in space. Its path is parametrized by the vector-valued function **r**(t) = 2**i** cos t + 2**j** sin t + t**k**. Find the velocity and acceleration vectors.

Answer: **v**(t) = <-2 sin t, 2 cos t, 1>
a(t) = <-2 cos t, -2 sin t, 0>

8) A particle moves in space. Its position vector at time t is **r**(t) = t**i** + **j** sin t + **k** cos t. Find the velocity and acceleration vectors at time t = 0.

Answer: **v**(0) = <0, 1, 0>
a(0) = <0, 0, -1>

12.6 Curvature and Acceleration
Solve the problem.

1) Find the arc length of the curve given by x = cos 3t, y = sin 3t, z = 4t, from t = 0 to t = $\frac{\pi}{2}$.

Answer: $\frac{5\pi}{2}$

2) Find the curvature of x = 2t - 3, y = t² + 2t + 5 at the point where t = 0.

Answer: $\frac{1}{4\sqrt{2}}$

3) Find the unit tangent and normal vectors for the curve y = t², x = t at the point t = 1.

Answer: **T** = $\frac{\sqrt{5}}{5}$**i** + $\frac{2\sqrt{5}}{5}$**j**; **N** = $\frac{2\sqrt{5}}{5}$**i** - $\frac{\sqrt{5}}{5}$**j**

4) Find the unit vectors **T** and **N** for the curve **r**(t) = cos (t)**i** + sin (t)**j** + 2t**k** at t = $\frac{\pi}{2}$.

Answer: **T** = $\frac{-\sqrt{5}}{5}$**i** + $\frac{2\sqrt{5}}{5}$**k**; **N** = $\frac{2\sqrt{5}}{5}$**i** + $\frac{\sqrt{5}}{5}$**k**

5) Find the arc length parameterization of the curve x(t) = 3cos t, y(t) = 3sin t, z(t) = 0, in terms of arc length s measured counterclockwise from the initial point (3, 0, 0).

Answer: x(s) = 3cos $\frac{s}{3}$; y(s) = -3sin $\frac{s}{3}$; z(s) = 0

6) Compute the speed (in miles/second) of the satellite at the nearest and farthest points of its orbit. Use $v = \frac{2\pi ab}{rT}$, a = 20,000mi, e = 0.5, and T = 24 hours.

Answer: nearest point = 1.260 mi/sec; farthest point = 1.454 mi/sec

7) A particle moves in space. Its position vector at time t is r(t) = 6ti + t²j + t³k. Find the tangential and normal components of its acceleration vector at time t = 1.

Answer: $a_T = \frac{22}{49}(6, 2, 3)$; $a_N = \frac{6}{49}(-22, 9, 38)$

8) Given: r(t) = 6ti + 3t²j + 2t³k, the trajectory of a point moving in space. Find the curvature of the trajectory, and the tangential and normal components of its acceleration vector at the point (6, 3, 2).

Answer: $k = \frac{\sqrt{2}}{18} \approx .0785674$

$a_T = \frac{18}{\sqrt{3}} \approx 10.392305$

$a_N = 6\sqrt{2} \approx 8.4852814$

9) Given: r(t) = i cos 2πt + j sin 2πt + tk, the trajectory of a point moving in space. Find the curvature, the unit tangent vector, and the principal unit normal vector at the point (1, 0, 1).

Answer: $k = \frac{4\pi^2}{1 + 4\pi^2} \approx .9752955$

$T(1) = \frac{1}{\sqrt{1 + 4\pi^2}}\langle 0, 2\pi, 1\rangle$

$N(1) = \langle -1, 0, 0\rangle$

10) Given: r(t) = ti + 4t²j + 2tk, the position vector of a particle moving in space. Find its tangential and normal components of acceleration, and the curvature of its trajectory at time t = 0.

Answer: $k = \frac{8}{5}$; $a_T = 0$; $a_N = 8$

11) A particle moves in space. At time t, its position is $x = t$, $y = t^2$, $z = \frac{4}{3}t^{3/2}$. Find the curvature of its trajectory, and the tangential and normal components of its acceleration at time t = 1.

Answer: $k = \frac{1}{9}$; $a_T = 2$; $a_N = 1$

12) Find the curvature of the Folium of Descartes with equation x³ + y³ = 3xy at the point (3/2, 3/2).

Answer: $k = 3\sqrt{3}$

13) Find the points on the graph of y = x² where the curvature is the greatest and where it is the least. Also find the curvature at all such points.

Answer: maximum of k = 2 at (0, 0); no minimum

14) One smooth parametrization of the ellipse with equation x² + 4y² = 4 is x = 2cos t, y = sin t, 0 ≤ t ≤ 2π. Use this parametrization to find the Cartesian coordinates of the points on the ellipse where its curvature is (a) maximal and (b) minimal.

Answer: (a) maximum at (2, 0) and (-2, 0)
(b) minimum at (0, 1) and (0, -1)

15) Find the point on the graph of y = eˣ at which the curvature is the greatest.

Answer: $\left(-\frac{1}{2}\ln 2, \frac{1}{\sqrt{2}}\right)$

16) Find the curvature of the parabola $y = x^2$ at $(0, 0)$.

Answer: $k = 2$

17) Does the graph of $y = \ln x$ have a point at which the curvature is minimal? Maximal?

Answer: no minimum, but maximum at $\left(\dfrac{1}{\sqrt{2}}, -\dfrac{1}{2}\ln 2\right)$

18) Find the point on the graph of $y = x^3$ at which the curvature is minimal.

Answer: $(0, 0)$

12.7 Cylinders and Quadric Surfaces
Solve the problem.

1) Describe the graph of $x + 2y - 3z = 6$.

Answer: a plane with intercepts $(6, 0, 0)$, $(0, 3, 0)$, and $(0, 0, -2)$

2) Describe the graph of $x^2 = 6(y^2 + z^2)$.

Answer: a circular cone with vertex at the origin, and x-axis as its axis

3) Describe the graph of $x^2 - 4y^2 + z^2 = 4$.

Answer: a hyperboloid of one sheet, y-axis as its axis, with traces in the xz-plane being larger circles and traces in the planes parallel to the y-axis being hyperbolas

4) Write the equation for the surface generated by revolving the curve $x = 4y^2$ about the x-axis. Then describe the surface.

Answer: equation $x = 4(y^2 + z^2)$; a paraboloid, x-axis as its axis, vertex at the origin

5) Write the equation for the surface generated by revolving the curve $x^2 - 2y^2 = 1$ about the y-axis. Then describe the surface.

Answer: equation $x^2 - 2y^2 + z^2 = 1$; a hyperboloid of one sheet, y-axis as its axis

6) Describe the traces of the curve $x^2 + 9z^2 = 9$ in the vertical planes (those parallel to the xz-plane).

Answer: ellipses with semiaxes 3 and 1

7) Describe the traces of the curve $z = 2(x^2 + y)$ in the horizontal planes (parallel to the xy-plane).

Answer: parabolas opening toward the negative y-axis

12.8 Cylindrical and Spherical Coordinates
Solve the problem.

1) Find the rectangular coordinates of the cylindrical point $(2, \dfrac{3\pi}{2}, 1)$.

Answer: $(0, -2, 1)$

2) Find the rectangular coordinates of the cylindrical point $(4, 0, \dfrac{\pi}{2})$.

Answer: $(0, 4, 0)$

3) Find both the cylindrical and spherical coordinates for the point with rectangular coordinates. (1, 1, 1).

   Answer: cylindrical: $(\sqrt{2}, \frac{\pi}{4}, 1)$ spherical: $(\sqrt{3}, \frac{\pi}{4}, \frac{\pi}{4})$

4) Describe the graph of r = 4.

   Answer: circular cylinder with radius 4 and axis of z

5) Describe the graph of $\theta = \frac{3\pi}{4}$.

   Answer: plane (y = -x)

6) Describe the graph of $r^2 - 2z^2 = 4$.

   Answer: hyperboloid of one sheet

7) Describe the graph of $\rho^2 - \rho^2 \cos\phi = 4$.

   Answer: circular cylinder with radius 2 and z-axis as the axis

8) Convert the equation x - y = z to both cylindrical and spherical coordinates.

   Answer: cylindrical: $r(\cos\theta - \sin\theta) = z$
   spherical: $\rho(\sin\phi\cos\theta - \sin\phi\sin\theta - \cos\phi) = 0$

9) Convert the equation $x^2 + y^2 + z^2 = 2xy$ to both cylindrical and spherical coordinates.

   Answer: cylindrical: $r^2(1 - 2\sin\theta\cos\theta) + z^2 = 0$
   spherical: $\rho^2 = 2\rho(\sin\phi)^2 \sin\theta\cos\theta$

10) Describe the surface giving by r = 2 for 0 ≤ z ≤ 2.

    Answer: the segment of a vertical circular cylinder of radius 2 between the plane z = 0 and z = 2

11) Describe the surface giving by $\rho = 1$ for $0 \leq \phi \leq \frac{\pi}{2}$.

    Answer: the upper $\frac{1}{2}$ of a sphere of radius 1

12) Describe the solid given by 0 ≤ r ≤ 2 and -1 ≤ z ≤ 2.

    Answer: the solid segment of a vertical circular cylinder of radius 2 between the planes z = -1 and z = 2

13) Describe the solid given by $0 \leq \phi \leq \frac{\pi}{4}$.

    Answer: the upper nappe of the cone whose sides are at 45 degree angles

14) The parabola $z = y^2$, x = 0 is rotated around the z-axis. Write a cylindrical-coordinate equation for the surface.

    Answer: $z = r^2$

15) Find the great circle distance in miles from New York (latitude 40.75° North, longitude 74° West) to Atlanta (latitude 33.75° North, longitude 84.4° West).

    Answer: 748 miles

Ch. 13 Partial Differentiation

13.1 Introduction

1) There are no exercises for this section.

    Answer:

13.2 Functions of Several Variables
Solve the problem.

1) State the largest possible domain of definition for the function $f(x, y) = \sqrt[3]{y^2 - x}$.

    Answer: the entire xy-plane

2) State the largest possible domain of definition for $f(x, y) = \dfrac{1 + \cos(x)}{xy}$.

    Answer: any one of the four quadrants, excluding the axes

3) State the largest possible domain of definition for $f(x, y, z) = \ln(4 + z^2 - x^2 - y^2)$.

    Answer: the region inside the hyperboloid of one sheet $4 = x^2 + y^2 - z^2$

4) Describe the graph of the function $f(x, y) = x^2 - y^2$.

    Answer: hyperbolic paraboloid

5) Describe the graph of the function $f(x, y) = 9 - x^2 - y^2$.

    Answer: a paraboloid opening downward

6) Describe the level surfaces of $f(x, y, z) = x^2 + y^2 + z^2 - 6x + 4y - 2z$.

    Answer: spheres centered at (3, -2, 1)

7) Describe the level surfaces of $f(x, y, z) = x^2 + z^2 + 16$.

    Answer: circular cylinders with y as an axis of symmetry

13.3 Limits and Continuity
Solve the problem.

1) Use the limit laws and consequences of continuity to evaluate $\lim\limits_{(x, y) \to (0, \frac{\pi}{2})} e^{2x}\sin(2x - y)$.

    Answer: -1

2) Use the limit laws and consequences of continuity to evaluate $\lim\limits_{(x, y, z) \to (1, 1, 1)} \dfrac{2x^2 - y^2 + z^2}{x + y - z}$.

    Answer: 2

3) Evaluate the limits $\lim\limits_{h \to 0} \dfrac{f(x + h, y) - f(x, y)}{h}$ and $\lim\limits_{k \to 0} \dfrac{f(x, y + k) - f(x, y)}{k}$ for $f(x, y) = 2x^2 - y^2$.

    Answer: $\lim\limits_{h \to 0} \dfrac{f(x + h, y) - f(x, y)}{h} = 4x$; $\lim\limits_{k \to 0} \dfrac{f(x, y + k) - f(x, y)}{k} = -2y$

4) Find the limit or state that it does not exist. $\lim_{(x, y, z) \to (0, 0, 0)} \frac{xy + xz + yz}{x^2 + y^2 + z^2}$.

   Answer: The limit does not exist.

5) Find the limit or state that it does not exist. $\lim_{(x, y) \to (0, 0)} e^{\sin(x + y - \pi/2)}$.

   Answer: e^{-1}

6) Determine the largest set of points in the xy-plane on which $f(x, y) = \ln(-4x^2 - y^2 + 1)$ defines a continuous function.

   Answer: inside the ellipse $4x^2 + y^2 = 1$

7) Determine the largest set of points in the xy-plane on which $f(x, y) = \sin^{-1}(x^2 - y)$ defines a continuous function.

   Answer: the region between the parabolas $y = x^2 - 1$ and $y = 1 + x^2$

8) Evaluate the limit $\lim_{(x, y) \to (0, 0)} \frac{x^3 y}{(x^2 + y^2)^{3/2}}$ by making the polar coordinate substitution $(x, y) = (r\cos(\theta), r\sin(\theta))$ and using the fact that $r \to 0$ as $(x, y) \to (0, 0)$.

   Answer: 0

9) Investigate the existence of the limit $\lim_{(x, y) \to (0, 0)} \frac{xy - y^2}{x^2 + y^2}$ by making the substitution $y = mx$.

   Answer: As the limit is approached along various lines $y = mx$ the result obtained is $\frac{m - m^2}{1 + m^2}$ which will vary depending on the value of m. Therefore, the limit does not exist.

10) Discuss the continuity of $g(x, y) = \begin{cases} \frac{\sin(2x^2 - y^2)}{2x^2 - y^2} & \text{unless } y^2 = 2x^2 \\ 1 & \text{if } y^2 = 2x^2 \end{cases}$.

    Answer: The function is continuous everywhere.

13.4 Partial Derivatives
Solve the problem.

1) Compute the first-order partial derivatives of $f(x, y) = \frac{2x}{x - y}$.

   Answer: $f_x = \frac{-2y}{(x - y)^2}$; $f_y = \frac{2x}{(x - y)^2}$

2) Compute the first-order partial derivatives of $f(x, y, z) = x^3 \sin(y) e^{z^2}$.

   Answer: $f_x = 3x^2 \sin(y) e^{z^2}$; $f_y = x^3 \cos(y) e^{z^2}$; $f_z = 2x^3 \sin(y) z e^{z^2}$

3) Compute the first-order partial derivatives of $f(u, v) = e^{(u + v)}(\cos(2u) + \sin(uv))$.

   Answer: $f_u = e^{(u + v)}(\cos(2u) - 2\sin(2u) + v\cos(uv) + \sin(uv))$
   $f_v = e^{(u + v)}(\cos(2u) + \sin(uv) + u\cos(uv))$

4) Verify that $z_{xy} = z_{yx}$ for $z = 5x^3 - 2yx^2 + 7x + 3x^2 - y^2x^3$.

   Answer: $z_{xy} = -4x - 6yx^2$; $z_{yx} = -4x - 6x^2y$; therefore $z_{xy} = z_{yx}$

5) Verify that $z_{xy} = z_{yx}$ for $z = e^{-5x^2}\sin(2y)$.

   Answer: $z_{xy} = -20x\,e^{-5x^2}\cos(2y)$; $z_{yx} = -20x\,e^{-5x^2}\cos(2y)$; therefore $z_{xy} = z_{yx}$

6) Find the equation of the plane tangent to $z = -\sin(\pi y x^2)$ at the point $P = (1, 1, 0)$.

   Answer: $z = 2\pi x + \pi y - 3\pi$

7) Find the equation of the plane tangent to $z = 2x^2 - y$ at the point $P = (1, 1, 1)$.

   Answer: $z = 4x - y - 2$

8) Determine whether there exists a function $f(x, y)$ having $f_x = 3yx^2 + 3y^2$ and $f_y = x^3 + 7 + 6xy$.

   Answer: yes, $f(x, y) = x^3 y + 3xy^2 + 7y + C$

9) Let $f(x, y, z) = x^2 y^2 z^2 + xy + xz + yz$, calculate the distinct second order partial derivatives of f and the third order partial derivative f_{xyz}.

   Answer: $f_{xx} = 2y^2z^2$; $f_{yy} = 2x^2z^2$; $f_{zz} = 2x^2y^2$

   $f_{xy} = 4xyz^2 + 1 = f_{yx}$; $f_{xz} = 4xy^2z + 1 = f_{zx}$; $f_{zy} = 4x^2yz + 1 = f_{yz}$

   $f_{xyz} = 8xyz$

10) A steady-state temperature function $u = u(x, y)$ for a thin flat plate satisfies Laplace's equation $\dfrac{\partial^2 u}{\partial x^2} + \dfrac{\partial^2 u}{\partial y^2} = 0$.

    Show that $u = e^{2x}\sin(2y)$ satisfies Laplace's equation.

    Answer: $\dfrac{\partial^2 u}{\partial x^2} = 4e^{2x}\sin(2y)$; $\dfrac{\partial^2 u}{\partial y^2} = -4e^{2x}\sin(2y)$; therefore $\dfrac{\partial^2 u}{\partial x^2} + \dfrac{\partial^2 u}{\partial y^2} = 0$

11) According to van der Waals' equation, 1 mole of a gas satisfies the equation $(p + \dfrac{a}{V^2})(V - b) = (82.06)T$ where p is the pressure in atmospheres, V is the volume in cubic centimeters, and T is the absolute temperature in Kelvins. For a particular gas, $a = 4 \times 10^6$, $b = 35.2$, and $V = 25000$ cm^3 when p is 1.1 atm and $T = 336.6$ Kelvins. Compute $\dfrac{\partial V}{\partial p}$ by differentiating van der Waals' equation with T held constant. Then estimate the change in volume that would result from an increase of 0.1 atm with T held at 336.6 kelvins.

    Answer: decrease of 0.109 cm^3

12) You are standing at the point where $x = y = 100$ on a hillside whose height (in feet above sea level) is given by $z = 50 + \dfrac{1}{1000}(x^2 + 2xy - 3y^2)$.

    (a) If you head due east, will you initially be ascending or descending? At what angle (in degrees) from the horizontal?

    (b) If you head due north, will you initially be ascending or descending? At what angle (in degrees) from the horizontal?

    Answer: (a) ascending; 21.8°
    (b) descending; −21.8°

13.5 Multivariable Optimization Problems
Solve the problem.

1) Find every point on the surface $z = f(x, y) = x^2 + y^2 - 4x + 8y - 7$ at which the tangent plane is horizontal.

   Answer: $(2, -4, -27)$

2) Find every point on the surface $z = f(x, y) = 5x^2 + 10x - 2y^3 + 3y^2 - 12$ at which the tangent plane is horizontal.

   Answer: $(-1, 0, -17)$ and $(-1, 1, -16)$

3) Find the highest point on $z = 4x - 2x^2 - 6y^2 - 3y^4$.

   Answer: $(1, 0, 2)$

4) Find the lowest point on $z = 4 + x^4 - 2x^3 + 3y^4 - 2y^3$.

   Answer: $(\frac{3}{2}, \frac{1}{2}, \frac{9}{4})$

5) Find the maximum and minimum values attained by $f(x, y) = x^2 - 2y^2 + 4y$ on R, where R is the triangle region with vertices $(0, 0)$, $(2, 0)$, and $(0, 2)$.

   Answer: minimum values: $f(0, 0) = 0$ and $f(0, 2) = 0$; maximum value: $f(0, 2) = 4$

6) Find the global maximum and minimum values of $f(x, y, z) = x^2 - 4xy + 2y^2 + 6yz + z^2 - 17$

   Answer: global minimum: $f(0, 0, 0) = -17$; no global maximum

7) A box with its base in the xy-plane has its four upper vertices on the surface with equation $z = 48 - 3x^2 - 4y^2$. What is the maximum possible volume of the box?

   Answer: $192\sqrt{3}$

8) A house in the form of a box is to have a volume of 10,000 ft^3. The walls admit heat at the rate of 5 units per minute per square foot, the roof at 3 units per minute per square foot, and the floor at 1 unit per minute per square foot. What should the shape of the house be to minimize the rate at which heat enters?

   Answer: base is a square of side $10 \cdot 5^{2/3}$ ft; height is $4 \cdot 5^{2/3}$ ft

9) The sum of three non negative numbers is 120. What is the maximum possible value of their product?

   Answer: $40^3 = 64,000$

10) A box in the first octant has three of its faces in the coordinate planes and the vertex opposite the origin in the plane with equation $x + 2y + 3z = 6$. What is the maximum possible volume of such a box?

    Answer: $\frac{4}{3}$ cubic units

13.6 Linear Approximations and Matrix Derivatives
Solve the problem.

1) Find the differential dw for $w = 2xy^2 - 3zy + 2zx^2$.

   Answer: $dw = (2y^2 + 4xz)dx + (4xy - 3z)dy + (2x^2 - 3y)dz$

2) Find the differential dw for $w = y\sin(x + z)$.

   Answer: $dw = y\cos(x+z)dx + \sin(x + z)dy + y\cos(x + y)dz$

3) Given $f(x, y, z) = \sqrt{x^2 + y^2 + z^2}$; $P(0, 3, 4)$, $Q(0.1, 2.9, 4.05)$, use the exact value $f(P)$ and the differential df to approximate the value $f(Q)$.

Answer: $f(P) = 5$; $f(Q) \approx 4.98$

4) Given $f(x, y) = (x + y)\sin(\frac{2\pi x}{y})$; $P(1, 4)$, $Q(1.05, 3.99)$, use the exact value $f(P)$ and the differential df to approximate the value $f(Q)$.

Answer: $f(P) = 5$; $f(Q) \approx 5.04$

5) Use differentials to approximate $\sqrt{24}(\sqrt[3]{26})(\sqrt[4]{83})$.

Answer: 43.822

6) A surveyor wants to find the area (in acres) of a certain field (1 acre is 43560 ft^2). She measures two different sides, finding them to be a = 450 ft and b = 625 ft, with a possible error of as much as 1 ft in each measurement. She finds the angle between these two sides to be 30°, with a possible error of as much as 0.5°. The field is triangular, so its area is given by $A = \frac{1}{2}ab \sin(\theta)$. Use differentials to estimate the maximum resulting error, in acres, in computing the area of the field by this formula.

Answer: 1.4 acres

7) The range of a projectile fired (in a vacuum) with initial velocity v_0 and inclination angle α from the horizontal is $R = \frac{1}{32}v_0^2\sin(2\alpha)$. Use differentials to approximate the change in range if v_0 is decreased from 400 to 395 ft/s and α is increased from 30° to 31°.

Answer: 7677 ft.

13.7 The Multivariable Chain Rule
Solve the problem.

1) Find $\frac{dw}{dt}$ both by using the chain rule and by expressing w explicitly as a function of t before differentiating.

$w = \sin(xyz)$, $x = t^2$, $y = 2t$, $z = t^3$

Answer: $\frac{dw}{dt} = 12t^5\cos(2t^6)$

2) Find $\frac{\partial r}{\partial x}$, $\frac{\partial r}{\partial y}$, and $\frac{\partial r}{\partial z}$ for $r = uv - u^2 - v^2 - w^2$; $u = 2x + z$, $v = x - y$, $w = y + z$.

Answer: $\frac{\partial r}{\partial x} = -6x - 3z$

$\frac{\partial r}{\partial y} = -3z - 4y$

$\frac{\partial r}{\partial z} = -3x - 3y - 4z$

3) Write the chain rule formulas giving the partial derivatives of p with respect to each of the independent variables for $p = f(u, v, w)$, $u = g(x, y)$, $v = h(y, z)$, $w = k(x, y, z)$

Answer: $\dfrac{\partial p}{\partial x} = \dfrac{\partial p}{\partial u}\dfrac{\partial u}{\partial x} + \dfrac{\partial p}{\partial w}\dfrac{\partial w}{\partial x}$

$\dfrac{\partial p}{\partial y} = \dfrac{\partial p}{\partial u}\dfrac{\partial u}{\partial y} + \dfrac{\partial p}{\partial v}\dfrac{\partial v}{\partial y} + \dfrac{\partial p}{\partial w}\dfrac{\partial w}{\partial y}$

$\dfrac{\partial p}{\partial z} = \dfrac{\partial p}{\partial v}\dfrac{\partial v}{\partial z} + \dfrac{\partial p}{\partial w}\dfrac{\partial w}{\partial z}$

4) Find $\dfrac{\partial z}{\partial x}$ and $\dfrac{\partial z}{\partial y}$ as functions of x, y, and z assuming that $z = f(x, y)$ satisfies the equation $xe^{yz} + ye^x + ze^y = 7$.

Answer: $\dfrac{\partial z}{\partial x} = \dfrac{-e^{yz} - ye^x}{e^y + xye^{yz}}$; $\dfrac{\partial z}{\partial y} = \dfrac{-xze^{yz} - e^x - ze^y}{xye^{yz} + e^y}$

5) Use implicit differentiation to find $\dfrac{\partial w}{\partial x}$ and $\dfrac{\partial w}{\partial y}$ for $w = \sqrt{(u+v)(x+y)}$, $u = x - y$, $v = x + y$

Answer: $\dfrac{\partial w}{\partial x} = \dfrac{2x + y}{\sqrt{2x(x+y)}}$; $\dfrac{\partial w}{\partial y} = \dfrac{x}{\sqrt{2x(x+y)}}$

6) Write an equation for the plane tangent at P(1, -1, 2) to the surface $x^2 - y^2 + 2z^2 = 8$.

Answer: $x + y + 4z = 8$

7) The sun is melting a rectangular block of ice. When the block's height is 1.5 ft, its length is 2 ft and its width is 4 ft each of its dimensions is decreasing at 2 inches per hour. What is the block's rate of change of volume V at that instant?

Answer: $-34 \text{ in}^3/\text{h}$

13.8 Directional Derivatives and the Gradient Vector
Solve the problem.

1) Find the gradient vector ∇f for $f(x, y, z) = x^2 - 2xy + 3z^2$ at the point P(1, 3, 2).

Answer: <-4, -2, 12>

2) Find the directional derivative of $f(x, y, z) = 2xy - xz + 3yz$ at P(1, 1, 1) in the direction of **v** = <2, -1, 1>.

Answer: $-\dfrac{\sqrt{6}}{6}$

3) Find the maximum directional derivative of $f(x, y, z) = 2x^2 - 3y^2 + 2z^2$ at the point P(1, 2, 1) and the direction in which it occurs.

Answer: maximum: $4\sqrt{11}$; direction: <1, -3, 1>

4) Suppose that the temperature at the point (x, y, z) in space with distance measured in kilometers, is given by $w = f(x, y, z) = xy + 2xz + yz - 10$ (in degrees celsius). Find the rate of change (in degrees Celsius per km) of the temperature at the point P(1, 1, 0) in the direction of the vector **v** = <1, 1, -1>

Answer: $\dfrac{2\sqrt{3}}{3}$

5) Suppose that the temperature w (in degrees Celsius) at the point (x, y) is given by
w = f(x, y) = 5 + 0.002x^2 + 0.003y^2 In what direction **u** should a grasshopper hop from the point (10, 20) in order to get warmer as quickly as possible? Find the directional derivative in this optimal direction.

Answer: direction: <1, 3>; directional derivative: 0.4

6) Write a Cartesian equation for the plane tangent to the graph of z = 2x^2 - y^2 at the point P(2, 1, 7).

Answer: 8x - 2y - z = 7

7) Given f(x, y, z) = x^{2}y^{3}z^6, in what direction is f(x, y, z) increasing the most rapidly at the point P(1, -1, 1)? What is its rate of increase in that direction?

Answer: increasing at a maximum rate in the direction of <-2, 3, -6>; maximum rate is 7

8) Let f(x, y, z) = x^{3}y^2. (a) What is the largest directional derivative of f at (1, 1)? (b) In what direction is the derivative in part (a)? (c) Find the directional derivative of f at (1, 1) in the direction -**i** -**j**. (d) What is the equation of the level curve of f that passes through the point (1, 1)? (e) Find the slope of the curve in part (d) at (1, 1). (f) Show that the gradient of f evaluated at (1, 1) is perpendicular to the tangent line to the level curve there.

Answer: (a) 3**i** + 2**j**; maximum value is $\sqrt{13}$

(b) $\dfrac{(3\mathbf{i} + 2\mathbf{j})}{\sqrt{13}}$

(c) $\dfrac{-5\sqrt{2}}{2}$

(d) x^{3}y^2 = 1 is the level curve

(e) slope = $-\dfrac{3}{2}$

(f) ∇f = 3**i** + 2**j**; 2**i** - 3**j** is parallel to the tangent at (1, 1), hence the two are perpendicular

9) Find a unit vector perpendicular at the point (1, 1, 2) to the surface with equation z = x^2 + y^2.

Answer: $\dfrac{2}{3}$**i** + $\dfrac{2}{3}$**j** - $\dfrac{1}{3}$**k** or its negative

10) Given f(x, y, z) = x^2 - 2yz, (a) what is the direction in which f is increasing most rapidly at the point (1, 2, 3)? (b) What is the directional derivative of f in that direction? (c) What is the directional derivative of f at the point (1, 2, 3) in the direction of 2**i** + 2**j** + **k**?

Answer: (a) f is increasing most rapidly in the direction $\left\langle \dfrac{1}{\sqrt{14}}, -\dfrac{3}{\sqrt{14}}, -\dfrac{2}{\sqrt{14}} \right\rangle$

(b) $2\sqrt{14}$

(c) $\dfrac{4}{3}$**i** - 4**j** - $\dfrac{4}{3}$**k**

11) Write a Cartesian equation for the plane tangent to the graph of z = x^2 - 4y^2 at the point (5, 2, 9).

Answer: 10x - 16y - z = 9

12) Given f(x, y, z) = x^3 - 3xyz + z^4, (a) in what direction is f increasing the most rapidly at the point (1, -1, 1)? (b) What is the rate of increase of f in that direction at that point?

Answer: (a) increasing most rapidly in the direction $\left\langle \dfrac{6}{\sqrt{94}}, -\dfrac{3}{\sqrt{94}}, \dfrac{7}{\sqrt{94}} \right\rangle$

(b) rate of increase: $\sqrt{94}$

13) Given $g(x, y, z) = xy - z^2$, (a) in what direction is g increasing the most rapidly at the point (1, 2, 1)? (b) What is the rate of increase of g in that direction at that point?

Answer: (a) increasing most rapidly in the direction $<\frac{2}{3}, \frac{1}{3}, -\frac{2}{3}>$

(b) maximum rate is 3

14) Write a Cartesian equation for the plane tangent to the graph of $z = 4x^2 - y^2$ at the point (1, 1, 3).

Answer: $8x - 2y - z = 3$

15) Write a Cartesian equation for the plane tangent to the graph of $z = 3x^2 - 4y^2$ at the point (2, 1, 8).

Answer: $12x - 8y - z = 8$

16) Find the points at which a plane parallel to $z = 2x + 2y$ can be tangent to the sphere with equation $x^2 + y^2 + z^2 = 9$.

Answer: (2, 2, -1), (-2, -2, 1)

17) Write an equation for the plane that is simultaneously tangent to the elliptic paraboloid with equation $z = 2x^2 + 3y^2$ and parallel to the plane with equation $4x - 3y - z = 10$.

Answer: $4x - 3y - z = \frac{11}{4}$

13.9 Lagrange Multipliers and Constrained Optimization
Solve the problem.

1) Use the method of Lagrange multipliers to find the extreme values of $3x - 4y + 12z$ on the spherical surface with equation $x^2 + y^2 + z^2 = 1$.

Answer: 13, -13

2) An aquarium has a rectangular slate bottom that costs 8 cents/in^2 and glass sides that cost 1 cent/in^2. The aquarium is to have volume 1024 in^3. Use the method of Lagrange multipliers to find the dimensions of the least expensive such aquarium.

Answer: base = $\sqrt[3]{256}$ in^2; height = $\sqrt[3]{16,384}$ in

3) Use the method of Lagrange multipliers to find the extreme values of the expression $x^2y^2z^2$ on the spherical surface with equation $x^2 + y^2 + z^2 = 3$ and the points on the surface at which these extrema occur.

Answer: minimum: 0 on the three circles:

$x = 0, y^2 + z^2 = 3$
$y = 0, x^2 + z^2 = 3$
$z = 0, x^2 + y^2 = 3$

maximum: 1 at all 8 points where $x^2 = y^2 = z^2 = 1$

4) Use the method of Lagrange multipliers to find the point or points on the graph of $z = 1 + xy$ closest to the origin.

Answer: (0, 0, 1)

5) Use the method of Lagrange multipliers to find the maximum possible volume of a right circular cone inscribed in a sphere of radius R.

Answer: $\dfrac{32\pi}{81}R^3$

6) Use the method of Lagrange multipliers to find the maximum possible volume of a cone inscribed in an inverted cone of radius R and height H.

Answer: $\dfrac{4\pi R^2 H}{81}$

7) A wire of length 120 cm is cut into three or fewer pieces which are then formed into a like number of circles. Use the method of Lagrange multipliers to maximize and minimize the total area of the circles.

Answer: maximum area: $\dfrac{3600}{\pi}$; minimum area: $\dfrac{1200}{\pi}$

8) Find the extreme values of the expression $E = xy + z^2$ on the spherical surface $x^2 + y^2 + z^2 = R^2$.

Answer: $E_{min} = \dfrac{R^2}{2}$; $E_{max} = -R^2$

9) Use the method of Lagrange multipliers to find the points of the ellipse $x^2 - xy + y^2 = 4$ closest to, and farthest from, the origin.

Answer: closest points: $(\dfrac{2}{\sqrt{3}}, -\dfrac{2}{\sqrt{3}})$, $(-\dfrac{2}{\sqrt{3}}, \dfrac{2}{\sqrt{3}})$; farthest points $(2, 2)$, $(-2, -2)$

10) The plane $2y + 4z = 5$ meets the cone $z^2 = 4(x^2 + y^2)$ in a curve. Use the method of Lagrange multipliers to find the point on this curve nearest to the origin.

Answer: $(0, \dfrac{1}{2}, 1)$

11) Use the method of Lagrange multipliers to find the point or points on the surface $x^2 + y^2 + 3xy + 4z^2 = 9$ nearest to the origin.

Answer: $(0, 0, \pm\dfrac{3}{2})$

12) A rectangular box has its lower four vertices in the xy-plane and its upper four on the ellipsoidal surface with equation $6x^2 + 3y^2 + 2z^2 = 450$. Use the method of Lagrange multipliers to find the maximum possible volume that such a box can have.

Answer: $V_{max} = 500\sqrt{6}$

13) Use the method of Lagrange multipliers to find the maximum value of $x + 2y + 2z$ on the spherical surface with equation $x^2 + y^2 + z^2 = 1$.

Answer: 3

14) Find the global maximum value and the global minimum value of the expression $x^2 + y^2 - z^2$ on the spherical ball satisfying the inequality $x^2 + y^2 + z^2 \le 1$.

Answer: maximum: 1; minimum: -1

15) Use the method of Lagrange multipliers to find the point on the intersection of $x^2 - xy + y^2 - z^2 = 1$ and $x^2 + y^2 = 1$ that is nearest the origin.

Answer: $(1, 0, 0), (-1, 0, 0), (0, 1, 0), (0, -1, 0)$

16) Use the method of Lagrange multipliers to find the maximum value of the expression $x - 2y + 2z$ on the spherical surface with equation $x^2 + y^2 + z^2 = 9$.

Answer: maximum: 9

17) Use the method of Lagrange multipliers to find the maximum value of the expression $x^2 - y^2$ on the spherical surface with equation $x^2 + y^2 + z^2 = 9$.

Answer: maximum: 9

13.10 Critical Points of Multivariable Functions
Solve the problem.

1) Locate and classify the critical points of $f(x, y) = x^4 + 4xy + y^4$.

Answer: $(0, 0, 0)$ is a saddle point; minima at $(-1, 1, -2)$ and $(1, -1, 2)$

2) Locate and classify the critical points of the function $f(x, y) = 4xy - x^4 - y^4 + 16$.

Answer: $(0, 0, 16)$ is a saddle point; maxima at $(1, 1, 18)$ and $(-1, -1, 18)$

3) Locate and classify the critical points of the function $g(x, y) = y^3 - 3x^2y$.

Answer: no extrema; $(0, 0, 0)$ is a saddle point

4) Locate and classify the critical points of the function $f(x, y) = x^3 - y^3 + x^2 + y^2$.

Answer: local minimum at $(0, 0, 0)$; local maximum at $(-\frac{2}{3}, \frac{2}{3}, \frac{8}{27})$

5) Locate and classify the critical points of the function $f(x, y) = x^2 - 2xy + y^3 - y$.

Answer: minimum at $(1, 1, -1)$; $(-\frac{1}{3}, -\frac{1}{3}, \frac{5}{27})$ is a saddle point

6) Locate and classify the critical points of the function $h(x, y) = x^2 - 4x + 4xy + y^2 - 16y$.

Answer: no extrema; $(\frac{14}{3}, -\frac{4}{3}, \frac{4}{3})$ is a saddle point

7) Locate and classify the critical points of the function $f(x, y) = x^4 - 4xy^3 + y^4$.

Answer: no extrema; $(0, 0, 0)$ is a saddle point

8) Locate and classify the critical points of the function $h(x, y) = x^2 + 2x + 3y^2 + 4y + 1$.

Answer: $(-1, -\frac{2}{3}, -\frac{4}{3})$ minimum

9) Locate and classify the critical points of the function $f(x, y) = x^3 - 3xy + y^3$.

Answer: $(1, 1, -1)$ is a local minimum; $(0, 0, 0)$ is a saddle point

10) Locate and classify the critical points of the function $g(x, y) = x^2 + xy$.

Answer: $(0, 0, 0)$ is a saddle point

11) Locate and classify the critical points of the function $g(x, y) = 2x^2 - 2xy + y^2 + 1$.

   Answer: $(0, 0, 1)$ is a minimum

12) Locate and classify the critical points of the function $g(x, y) = x^3 + 6xy - y^3$.

   Answer: $(0, 0, 0)$ is a saddle point; $(2, -2, -8)$ is a minimum

13) Let $f(x, y)$ denote the square of the distance from $(0, 1, 0)$ to a typical point on the surface $z = \sqrt{x + y}$. Find and classify the critical points of f.

   Answer: $(-\frac{1}{2}, \frac{1}{2}, 0)$ local minimum

14) Find and classify the critical points of $f(x, y) = \cos(\frac{\pi x}{2}) \cos(\frac{\pi y}{2})$.

   Answer: The critical points are all ordered pairs (x, y) such that either both are even integers or both are odd integers. If both are even, then the point is a minimum if both are divisible by 4, otherwise it is a maximum. If both are odd, then if both are of the form 4k + 1 or 4k+3, it can not be determined by the 2-variable second derivative test, otherwise the point is a saddle point.

15) Let $f(x, y) = \dfrac{x^2 y^2}{x^2 + y^2}$, classify the behavior of f near the critical point $(0, 0)$.

   Answer: The critical point $(0, 0)$ is a local minimum as are all points where $x = 0$ or $y = 0$. The surface increases over the domain region above the four quadrants.

Ch. 14 Multiple Integrals

14.1 Double Integrals

Solve the problem.

1) Calculate the Riemann sum for $\iint x^2 + y \, dA$ where $R = [0,2] \times [0,2]$ using the partition of four unit squares where each (x_i^*, y_i^*) is the center point of the ith rectangle.

   Answer: 9

2) Evaluate $\int_1^3 \int_0^1 (2x - 3y) \, dx \, dy$.

   Answer: −10

3) Evaluate $\int_{-1}^{2} \int_{-1}^{1} (2xy + 3x^2y^2) \, dy \, dx$.

   Answer: 6

4) Evaluate $\int_1^3 \int_{-2}^{-1} (x^2y - 2xy^3) \, dy \, dx$.

   Answer: 17

5) Evaluate $\int_0^{\pi/4} \int_0^2 e^{2x} \cos(y) \, dx \, dy$.

   Answer: $\dfrac{\sqrt{2}(e^4 - 1)}{4}$

6) Evaluate $\int_0^1 \int_1^2 \dfrac{1}{x^2y + y} \, dy \, dx$.

   Answer: $\dfrac{\pi}{4}\ln(2)$

14.2 Double Integrals Over More General Regions

Solve the problem.

1) Evaluate the iterated integral $\int_0^2 \int_0^x (x + 2y) \, dy \, dx$.

   Answer: $\dfrac{16}{3}$

2) Evaluate the iterated integral $\int_0^1 \int_0^{\sqrt{x}} (x - 2y - 7) \, dy \, dx$.

   Answer: $-\dfrac{143}{30}$

3) Evaluate the iterated integral $\int_{-1}^{2} \int_{0}^{1/x^2} x^3 e^{x^3 y} \, dy \, dx$.

Answer: $e^2 - 3 - e^{-1} = \dfrac{e^3 - 3e - 1}{e}$

4) Evaluate the integral of $f(x, y) = x^2 y$ over the plane region bounded by the parabola $y = -x^2$ and the line $y = -4$.

Answer: $-\dfrac{512}{21}$

5) Evaluate the integral of $f(x,y) = 2x$ over the plane region in the first quadrant bounded by $y = \cos(x)$.

Answer: $\pi - 2$

6) Evaluate the integral of $f(x,y) = x - 2y$ over the plane region R, where R is the triangle with vertices (0, 0), (0, 2), and (2, 0).

Answer: $-\dfrac{4}{3}$

7) Find the approximate value of the integral $\iint_R x \, dA$ where R is the region between the curves $y = x^3 + 2$ and $y = 3x^2$. Use a calculator to approximate the points of intersection.

Answer: 4.5

8) Find the approximate value of the integral $\iint_R x \, dA$ where R is the region bounded by the curves $y = x^2 - 1$ and $y = \sin(x)$. Use a calculator to approximate the points of intersection.

Answer: 0.6025

9) For $\iint_R x^2 y^2 \, dA$ where R is the square with vertices (± 1, 0) and (0, ± 1). Use the symmetry of the region around the coordinate axes to reduce the labor of the evaluation of the integral

Answer: $\dfrac{4}{9}$

10) Evaluate $\int_{0}^{2} \int_{x/2}^{1} e^{-y^2} \, dy \, dx$ by first reversing the order of integration.

Answer: $1 - e^{-1}$

11) Evaluate $\int_{0}^{\pi/2} \int_{y}^{\pi/2} \dfrac{\sin x}{x} \, dx \, dy$ by first reversing the order of integration.

Answer: $\int_{0}^{\pi/2} \int_{0}^{x} \dfrac{\sin x}{x} \, dy \, dx = 1$

12) Evaluate $\displaystyle\int_0^1 \int_{\sqrt{y}}^1 \frac{3}{x^3+1}\,dx\,dy$.

Answer: $\ln(2)$

13) Reverse the order of integration, then evaluate $\displaystyle\int_0^1 \int_y^1 x^2 y\,dx\,dy$.

Answer: $\displaystyle\int_0^1 \int_0^x x^2 y\,dy\,dx = \frac{1}{10}$

14) Correctly reverse the order of integration, then evaluate $\displaystyle\int_0^1 \int_y^1 x e^y\,dx\,dy$.

Answer: $\displaystyle\int_0^1 \int_0^x x e^y\,dy\,dx = \frac{1}{2}$

15) Reverse the order of integration, then evaluate $\displaystyle\int_0^1 \int_0^x x^2 y^2\,dy\,dx$.

Answer: $\displaystyle\int_0^1 \int_y^1 x^2 y^2\,dx\,dy = \frac{1}{18}$

16) Reverse the order of integration, but DO NOT evaluate $\displaystyle\int_0^4 \int_{\sqrt{y}}^{y/2} e^x \cos y\,dx\,dy$.

Answer: $\displaystyle\int_0^2 \int_{2x}^{x^2} e^x \cos y\,dy\,dx$

17) Reverse the order of integration, then evaluate $\displaystyle\int_0^1 \int_{x^2}^{x^{1/3}} (x+y)\,dy\,dx$.

Answer: $\displaystyle\int_0^1 \int_{y^3}^{\sqrt{y}} (x+y)\,dx\,dy = \frac{53}{140}$

18) Reverse the order of integration, then evaluate $\int_0^1 \int_{-\sqrt{1-y^2}}^{\sqrt{1-y^2}} x^3 y^3 \, dx \, dy$.

Answer: $\int_{-1}^{1} \int_0^{\sqrt{1-x^2}} x^3 y^3 \, dy \, dx = 0$

14.3 Area and Volume by Double Integration
Solve the problem.

1) Use a double integral to find the area of the region in the xy-plane bounded by $y = 5x$ and $y = x^2 - 6$.

   Answer: $\dfrac{343}{6}$

2) Use a double integral to find the area of the region in the xy-plane bounded by $y = x^2 - 1$ and $y = \dfrac{x^2}{2} + 1$.

   Answer: $\dfrac{16}{3}$

3) Find the volume of the solid that lies below $z = 3x^2 y^2$ and above the region in the xy-plane bounded by $y = x^2$ and $y = 4$.

   Answer: $\dfrac{2048}{9}$

4) Find the volume of the solid that lies below $z = x^2 + y^2$ and above the region in the xy-plane bounded by $x = 0$, $y = 0$, and $y = 4 - 2x$.

   Answer: $\dfrac{40}{3}$

5) Find the volume of the solid bounded by the planes $x = 0$, $y = 0$, $z = 0$, and $4x + 3y + 2z = 12$.

   Answer: 12

6) Find the volume of a three sided pyramid by double integration. The pyramid has length down a slant edge of L. Think of the pyramid lying on one of its sides with its vertex at the origin and the other two sides along the other two coordinate planes.

   Answer: $\dfrac{L^3}{6}$

7) A solid has its base in the xy-plane; the base is the part of the circular disk $x^2 + y^2 \leq 1$ that lies in the first quadrant. The solid has vertical sides and its height at the point (x, y) is $x + y$. Find the volume of the solid by means of a double integral in Cartesian coordinates.

   Answer: $\dfrac{2}{3}$

Calculus 286

8) Use a double integral in Cartesian coordinates to find the volume of the tetrahedron in space with vertices at (2, 0, 0), (0, 1, 0), (0, 0, 4), and (0, 0, 0).

   Answer: $\dfrac{4}{3}$

9) The plane region R is bounded by the graphs of $y = x$ and $y = x^2$. Find the volume over R and beneath the graph of $f(x, y) = x + y$.

   Answer: $\dfrac{3}{20}$

10) Find the volume of the solid between the xy-plane and the graph of $z = xe^y$, $0 \le x \le 1$, $0 \le y \le 1$.

    Answer: $\dfrac{e}{2} - \dfrac{1}{2}$

11) A solid has its base in the xy-plane and is bounded by the graphs of $y = x^2$ and $x = y^2$ for $0 \le x \le 1$ and $0 \le y \le 1$. Its height at the point (x, y) is xy. Find its volume.

    Answer: $\dfrac{1}{12}$

12) The base of a solid is the region bounded by graphs of $y = x^2$ and $y = x$ in the xy-plane. It has vertical sides and is bounded above by the surface with equation $z = x - y$. Find its volume.

    Answer: $\dfrac{1}{60}$

14.4 Double Integrals in Polar Coordinates
Solve the problem.

1) Find the area bounded by the circle $r = 2$ using a double integral in polar coordinates.

   Answer: 4π

2) Use a double integral in polar coordinates to find the area outside $r = 2$ and inside $r = 4\cos(\theta)$.

   Answer: $\sqrt{3} + \dfrac{2\pi}{3}$

3) Use double integration in polar coordinates to find the volume of the solid that lies below $z = x^2 + y^2$ and above the plane region inside $r = 2$

   Answer: 4π

4) Evaluate the double integral $\displaystyle\int_{-3}^{3}\int_{0}^{\sqrt{9-x^2}} e^{(x^2 + y^2)}\, dy\, dx$ by first converting it to polar coordinates.

   Answer: $\dfrac{\pi}{2}(e^9 - 1)$

5) Find the volume of the solid bounded below by $z = -2 + x + y$ and above by $z = 5$, that lies above the plane region inside $r = 1$.

   Answer: 8π

6) Find the volume of the upper hemisphere of a sphere of radius a.

   Answer: $\dfrac{2\pi a^3}{3}$

7) Find the volume of the "ice cream cone" bounded by the sphere $x^2 + y^2 + z^2 = 1$ and the cone $z = \sqrt{x^2 + y^2 - 1}$.

   Answer: π

8) Use a double integral in polar coordinates to find the volume of the solid bounded by the elliptical paraboloids $z = x^2 + y^2$ and $z = 6 - (x^2 + 2y^2)$.

   Answer: 6π

9) The region R in the xy-plane is bounded by semicircle $x^2 + y^2 = a^2$, $y \geq 0$, and by the straight line segment from $(-a, 0)$ to $(a, 0)$. Find the volume of the solid with base R, vertical sides, and bounded above by the graphs of the equation $z = (x^2 + y^2)^{3/2}$.

   Answer: $\dfrac{\pi a^5}{5}$

10) Find the volume of the solid bounded above by the graph of $z = 1 + xy$, with vertical sides, and with base the circular disk $x^2 + y^2 \leq 1$ in the xy-plane.

    Answer: π

11) Find the volume of the solid bounded below by the xy-plane and above by the graph of $x^2 + y^2 + z = 1$.

    Answer: $\dfrac{\pi}{2}$

12) Correctly convert to a double integral in polar coordinates, the iterated Cartesian integral

    $$\int_0^1 \int_0^{\sqrt{1-y^2}} \frac{1}{1 + x^2 + y^2} \, dx \, dy,$$ then evaluate the polar double integral.

    Answer: $\displaystyle\int_0^1 \int_0^{\pi/2} \frac{1}{1+r^2} \, d\theta \, dr = \frac{1}{4} \ln 2\pi$

14.5 Applications of Double Integrals
Solve the problem.

1) Find the centroid of the plane region bounded by $x = 1$, $x = 3$, $y = 0$, and $y = 4$. Assume density $\delta \equiv 1$.

   Answer: $(2, 2)$

2) Find the centroid of the plane region bounded by $x = 0$, $x = 2$, $y = 0$, and $y = x^2 + 1$

   Answer: $\left(\dfrac{9}{7}, \dfrac{103}{70}\right)$

3) Find the mass and the centroid of the plane lamina with density $\delta(x, y) = x + 2y$ over the triangular region bounded by $x = 0$, $y = 0$, and $y = 2 - x$.

   Answer: mass = 4; centroid at $\left(\dfrac{2}{3}, \dfrac{5}{6}\right)$

4) Find the mass and centroid of the plane lamina with density $\delta(x, y) = x^2 y$ bounded by the parabolas $y = x^2$ and $x = y^2$.

   Answer: mass $= \dfrac{3}{56}$; centroid at $\left(\dfrac{7}{10}, \dfrac{56}{81}\right)$

5) Find the mass and centroid of the plane lamina with $\delta(x, y) = 2x + y$ over the rectangle with vertices $(0, 0)$, $(0, b)$, $(a, 0)$, and (a, b)

   Answer: mass $= ab(a + \dfrac{b}{2})$; centroid at $\left(\dfrac{8a^2 + 3ab}{12a + 6b}, \dfrac{3ab + 2b^2}{6a + 3b}\right)$

6) Find the mass and centroid of the plane lamina with $\delta(x, y) = \sqrt{x^2 + y^2}$ over the region inside $r = 2\cos(\theta)$ and outside $r = 1$.

   Answer: mass $= 2\sqrt{3} - \dfrac{2\pi}{9}$; centroid at $(\approx 1.3776, 0)$

7) Find the polar moment of inertia I_0 of the region bounded by $r = 5$ with $\delta(x, y) = r^2$.

   Answer: $\dfrac{15625\pi}{3}$

8) Find the radii of gyration $\hat{x}$ and $\hat{y}$ around the coordinate axes of the lamina above the triangular region with vertices $(0, 0)$, $(0, 1)$, and $(1, 0)$ and with $\delta(x, y) = 2x + y$.

   Answer: $\hat{x} = \dfrac{\sqrt{210}}{30}$; $\hat{y} = \dfrac{\sqrt{6}}{6}$

9) Find the centroid of the portion of the cardioid $y = 1 + \cos(\theta)$ in the first quadrant. Use $\delta \equiv 1$.

   Answer: $\left(\dfrac{5\pi + 16}{6\pi + 16}, \dfrac{10}{3\pi + 8}\right)$

10) Find the centroid of a plane region consisting of a semicircle of radius a sitting atop a rectangular region of width $2a$ and height a, that has the x-axis as its lower edge and is centered at the origin. Then apply the first theorem of Pappus to find the volume generated by rotating this region around the x-axis.

    Answer: centroid at $\left(0, \dfrac{10a^3 + 3\pi a^3}{12a^2 + 3\pi a^2}\right)$; volume $= \dfrac{5\pi}{3} a^3$

11) Find the mass and centroid of the plane lamina bounded by the polar equation $r = 2\cos(\theta)$ that has density function $\delta(x, y) = x$.

    Answer: $\left(\dfrac{5}{4}, 0\right)$

12) Find the centroid of the plane lamina of constant density in the shape of the quarter-circle of radius 4 in the first quadrant.

    Answer: $\left(\dfrac{16}{3\pi}, \dfrac{16}{3\pi}\right)$

13) A lamina in the xy-plane is a square with vertices at $(1, 1)$, $(1, -1)$, $(-1, -1)$, and $(-1, 1)$ and has constant density 1. Find its moment of inertia about (a) the z-axis and (b) the x-axis.

    Answer: (a) $\dfrac{8}{3}$; (b) $\dfrac{4}{3}$

14.6 Triple Integrals
Solve the problem.

1) Compute the value of the triple integral for $f(x, y, z) = x^2 + y - z$ over the rectangular box $0 < x < 1$, $1 < y < 3$, $-1 < z < 0$.

   Answer: $\dfrac{17}{3}$

2) Compute the value of the triple integral for $f(x, y, z) = x + 2y$ where the solid is the tetrahedron bounded by the coordinate planes and the first octant part of the plane with equation $2x + y + 4z = 8$.

   Answer: $\dfrac{160}{3}$

3) Find the volume of the solid bounded by $x = 0$, $y = 0$, $z = 0$, and $x + 2y + 3z = 6$ by triple integration.

   Answer: 18

4) Find the volume of the solid bounded by $x = 0$, $y = 0$, $z = 0$, and $4x + 3y + 2z = 12$ by triple integration.

   Answer: 12

5) Find the centroid of the solid bounded by $x = 0$, $y = 0$, $z = 0$, and $4x + 3y + 2z = 12$, assume $\delta(x, y, z) \equiv 1$.

   Answer: $(\dfrac{3}{4}, 1, \dfrac{3}{2})$

6) Find the moment of inertia around the z-axis of the solid bounded by $x = 0$, $y = 0$, $z = 0$, $y = 1 - x^2$, and $4x + 3y + 2z = 12$, assume $\delta(x, y, z) \equiv 1$.

   Answer: $\dfrac{527}{420}$

7) Find the centroid of the pyramid bounded by $x = 0$, $y = 0$, $z = 0$, and $z = 6 - 3x - 2y$ if $\delta(x, y, z) = x$.

   Answer: $\left(\dfrac{4}{5}, \dfrac{3}{5}, \dfrac{6}{5}\right)$

8) Find the average value of the density function $\delta(x, y, z) = x$ at the point of the pyramid given by $x = 0$, $y = 0$, $z = 0$, $x + 2y + 3z = 6$.

   Answer: $\dfrac{3}{2}$

9) Find the moment of inertia of right circular cone of radius R, height H, uniform density δ, and total mass M about its natural axis of symmetry.

   Answer: $\dfrac{3}{10}MR^2$

10) Find the centroid of the uniform solid eighth-sphere $x^2 + y^2 + z^2 \le R$, $x \ge 0$, $y \ge 0$, $z \ge 0$.

    Answer: $\left(\dfrac{3R}{8}, \dfrac{3R}{8}, \dfrac{3R}{8}\right)$

11) Set up an iterated triple integral, complete with correct limits of integration, for the volume V of the first octant pyramid bounded by the plane x + y + z = 3 and the three coordinate planes. Do not evaluate the integral.

Answer: $\int_0^3 \int_3^{3-x} \int_0^{3-x-y} dz\, dy\, dx$

12) Find the average distance of points of a circular disk of radius R from its center.

Answer: $\frac{2}{3}R$

13) Find the average distance of points of a square region of edge length a from one of its corner points.

Answer: $\frac{a}{3}[\sqrt{2} - \ln a + \ln(a + a\sqrt{2})]$

14.7 Integration in Cylindrical and Spherical Coordinates
Solve the problem.

1) Find the moment of inertia around the z-axis, using cylindrical coordinates, of a sphere of radius a assuming $\delta(x, y, z) = \sqrt{x^2 + y^2}$.

Answer: $\frac{\pi a^4}{2}$

2) Find the mass of a cylinder $0 \le r \le 5, 0 \le z \le 10$, if its density at (x, y, z) is $\sqrt{x^2 + y^2}$.

Answer: $\frac{2500\pi}{3}$

3) Find the centroid of a cylinder $0 \le r \le 5, 0 \le z \le 10$, if its density at (x, y, z) is $x^2 + y^2$.

Answer: (0, 0, 5)

4) Use spherical coordinates to find the mass and centroid of the solid hemisphere $x^2 + y^2 + z^2 \le a^2, z \ge 0$ if its density is proportional to the distance from the z-axis is $\delta(x, y, z) = k\sqrt{x^2 + y^2}$.

Answer: mass = $\frac{ka^4\pi^2}{8}$; centroid at $\left(0, 0, \frac{16a}{15\pi}\right)$

5) Use spherical coordinates to find the moment of inertia I_z of the upper hemisphere of a sphere of radius a.

Answer: $\frac{4a^5\pi}{15}$

6) Use spherical coordinates to find the mass and centroid of the fat "ice cream cone" $0 \le \theta \le 2\pi, 0 \le \phi \le \frac{\pi}{4}$, and $0 \le \rho \le 2a\cos(\phi)$, if its density at (x, y, z) is $\delta(x, y, z) = z$.

Answer: mass = $\frac{7a^4\pi}{6}$; centroid at $\left(0, 0, \frac{9a}{7}\right)$

7) Find, by a triple integral in spherical coordinates, the volume of the solid that is bounded above by the sphere $\rho = 4$ and below by the cone $\phi = \frac{\pi}{3}$.

Answer: $\frac{64\pi}{3}$

8) The spherical equation $\rho = a(1 + \cos \phi)$ does not involve θ, and thus its graph is a surface of revolution around the z-axis. (It resembles an inverted apple with the stem at the origin.) Find the volume it encloses by means of a multiple integral in spherical coordinates.

Answer: $\frac{8\pi a^3}{3}$

9) Find the average distance of points of a solid ball of radius R from its center.

Answer: $\frac{3}{4}R$

10) Find the average distance of points of a solid ball of radius R from a point P on its surface. [Suggestion: Use the ball whose surface has the spherical equation $\rho = 2R\cos \phi$ where $0 \le \phi \le \frac{\pi}{2}$ and $0 \le \theta \le 2\pi$; choose for P, the "south pole" of the ball, which is located at the origin.]

Answer: $\frac{6}{5}R$

11) A sphere of radius 2 has a circular hole of radius 1 drilled through its center (the center line of the hole lies of a diameter of the sphere). Find the volume of material removed by using a triple integral in cylindrical coordinates.

Answer: $\frac{4}{3}\pi(8 - 3^{3/2})$

12) Find the volume of the solid above the xy-plane inside both the cylinder $r = \cos \theta$ and the ellipsoid $z^2 + 4r^2 = 4$.

Answer: $\frac{2}{3}\pi - \frac{8}{9}$

13) Find, by means of a triple integral, the volume of the solid that is bounded above by the paraboloid $z = 4 - x^2 - y^2$ and which lies in the first octant.

Answer: 2π

14) Find the volume of the solid that lies inside both the sphere $x^2 + y^2 + z^2 = 4$ and the cylinder $x^2 + y^2 = 2x$.

Answer: changing to cylindrical coordinates yields $\frac{16}{3}\pi - \frac{64}{9}$

14.8 Surface Area
Solve the problem.

1) Find the area of the portion of the surface of the cone $z = 2\sqrt{x^2 + y^2}$ that lies above the square in the xy-plane defined by $-1 \le x \le 1, -1 \le y \le 1$.

Answer: $4\sqrt{5}$

2) A parameterization of a quadric surface is given by $x = au\cos(v)$, $y = bu\sin(v)$, and $z = cu$. Use identities such as $\cos^2 t + \sin^2 t = 1$ and $\cosh^2 t - \sinh^2 t = 1$ to identify the surface.

Answer: cone

3) Find the area of the part of the hemisphere $x^2 + y^2 + z^2 = a^2$, $z \geq 0$, that lies within the cylinder with cylindrical coordinates $r = a\cos\theta$.

Answer: $2a^2(\pi - 2)$

4) Find the area of the part of the surface $z = x^2 - y^2$, that lies within the cylinder with equation $x^2 + y^2 = 4$.

Answer: $\dfrac{\pi}{6}(17^{3/2} - 1)$

5) Find the area of the part of the paraboloid $z = a^2 - r^2$, that lies above the xy-plane.

Answer: $\dfrac{\pi}{6}[(4a^2 + 1)^{3/2} - 1]$

6) Use a surface integral to find the curved surface area of a right circular cylinder of radius R and height H.

Answer: $2\pi RH$

7) Use a surface integral in cylindrical coordinates to find the curved surface area of a right circular cone of radius R and height H.

Answer: $\pi R\sqrt{R^2 + H^2}$

8) Find the area of the part of the surface $2z = x^2$ that lies directly above the triangle in the xy-plane with vertices at (0, 0), (1, 0), and (1, 1).

Answer: $\displaystyle\int_0^1 \int_0^x \sqrt{1 + x^2}\, dx\, dy = \dfrac{2\sqrt{2} - 1}{3}$

9) Find the area of the part of the surface $z = x + y^2$ that lies directly above the triangle in the xy-plane with vertices at (0, 0), (1, 0), and (1, 1).

Answer: $\displaystyle\int_0^1 \int_0^x \sqrt{2 + 4y^2}\, dx\, dy = \dfrac{3\sinh^{-1} 2 + \sqrt{5} + 1}{12}$

10) Find the area of the elliptical region that is cut from the plane $2x + 3y + z = 6$ by the cylinder with equation $x^2 + y^2 = 2$.

Answer: $\displaystyle\int_{-\sqrt{2}}^{\sqrt{2}} \int_{-\sqrt{2-x^2}}^{\sqrt{2-x^2}} \sqrt{14}\, dy\, dx = \pi\sqrt{56}$

11) Find the area of the region that is cut from the surface $z = x^2 - y^2$ by the cylinder with equation $x^2 + y^2 = 1$.

Answer: $\displaystyle\int_{-1}^{1} \int_{-\sqrt{1-x^2}}^{\sqrt{1-x^2}} \sqrt{1 + 4x^2 + 4y^2}\, dy\, dx = \dfrac{\pi(5\sqrt{5} - 1)}{6}$

12) Find the area of the part of the sphere $x^2 + y^2 + z^2 = a^2$ that lies within the cylinder with equation $x^2 + y^2 = ay$.

Answer: $2\pi a^2$

13) Describe the surface that is defined by the equation $\rho = 2a \sin \phi$, and show that its total surface area is $A = 4\pi^2 a^2$.

Answer: A torus of radius a with center on the xy-plane at a distance a from the origin.

14.9 Change of Variables in Multiple Integrals
Solve the problem.

1) Solve for x and y in terms of u and v and then compute the Jacobian $\dfrac{\partial(x, y)}{\partial(u, v)}$ for $u = 2x + y$ and $v = x - y$.

Answer: $-\dfrac{1}{3}$

2) Solve for x and y in terms of u and v and then compute the Jacobian $\dfrac{\partial(x, y)}{\partial(u, v)}$ for $u = 4(x^2 + y^2)$ and $v = 2(x^2 - y^2)$.

Answer: $-\dfrac{1}{8(u^2 - 4v^2)}$

3) Let R be the parallelogram bounded by the lines $x + y = 2$, $x + y = 4$, $2x - y = 1$, and $2x - y = 4$. Use the transformation $u = x + y$ and $v = 2x - y$ to find the area of R.

Answer: 2

4) Let R be the first quadrant region bounded by the circles $x^2 + y^2 = 2x$ and $x^2 + y^2 = 4x$ and the circles $x^2 + y^2 = 2y$ and $x^2 + y^2 = 6y$. Use the transformation $u = \dfrac{2x}{x^2 + y^2}$ and $v = \dfrac{2y}{x^2 + y^2}$ to find the area of R.

Answer: $\dfrac{1}{2}$

5) Find the volume of the region in the first octant that is bounded by the hyperbolic cylinders $xy = 1$, $xy = 9$, $xz = 1$, $xz = 4$, $yz = 1$, and $yz = 16$. Use the transformation $u = xy$, $v = xz$, $w = yz$ Note that $uvw = x^2 y^2 z^2$.

Answer: 24

Ch. 15 Vector Calculus
15.1 Vector Fields
Solve the problem.

1) Calculate the divergence and curl of the vector field $F(x, y, z) = 2xi + 3yj + 4zk$

   Answer: 9; 0

2) Calculate the divergence and curl of the vector field $F(x, y, z) = xyzi - 4xzj + 3yzk$

   Answer: $3y + yz$; $<3z + 4x, -xy, -4z - xz>$

3) Calculate the divergence and curl of the vector field $F(x, y, z) = 3xy^2 i + 2y^2 zj + x^2 yzk$

   Answer: $3y^2 + 4yz + x^2 y$; $<x^2 z - 2y^2, 2xyz, -6xy>$

4) Calculate the divergence and curl of the vector field $F(x, y, z) = (e^{yz}\cos x)i - (e^{xz}\sin y)j$

   Answer: $-e^{yz}\sin x - e^{xz}\cos y$; $<-xe^{xz}\sin y, -ye^{yz}\cos x, -ze^{xz}\sin y - ze^{yz}\cos x>$

15.2 Line Integrals
Solve the problem.

1) Evaluate the line integrals $\int_C f(x, y)\, ds$, $\int_C f(x, y)\, dx$, and $\int_C f(x, y)\, dy$ along the parametric curve:

   $f(x, y) = 3x + 4y^2$; $x = 2t - 1$, $y = t + 5$, $-1 \le t \le 2$.

   Answer: $372\sqrt{5}$; 744; 372

2) Evaluate $\int_C P(x, y)\, dx + Q(x, y)\, dy$ given:

   $P(x, y) = y^2$, $Q(x, y) = 3x$; C is the part of the graph of $y = 3x^2$ from $(-1, 3)$ to $(2, 12)$.

   Answer: $\dfrac{567}{5}$

3) Evaluate the line integral $\int_C \mathbf{F} \cdot \mathbf{T}\, ds$ along the path C given:

   $F(x, y, z) = zi - xj + yk$; $x = \cos t$, $y = \sin t$, $z = 3t^2$, $0 \le t \le \dfrac{\pi}{2}$.

   Answer: $12 - \dfrac{13\pi}{4}$

4) Evaluate $\int_C f(x, y, z)\, ds$ along the path C given:

   $f(x, y, z) = 3x + 4xy + 2z$; C is the curve $x = 2t$, $y = t^2$, $z = 4t + 1$, $0 \le t \le 1$.

   Answer: ≈ 51.58

15.3 The Fundamental Theorem and Independence of Path
Solve the problem.

1) If the given vector field is conservative, find a potential function for the field.

   $F(x, y) = (8xy - 2y + 3y^2)i + (4x^2 - 2x + 6xy)j$

   Answer: conservative; $f(x, y) = 4x^2y - 2xy + 3xy^2$

2) If the given vector field is conservative, find a potential function for the field.

   $F(x, y) = (e^x + \frac{y}{x} - \sin x)i + (\ln x + 1)j$

   Answer: conservative; $f(x, y) = \cos x + y \ln x + e^x$

3) Given that the line integral is independent of path in the entire xy-plane, calculate the value of the line integral:

   $$\int_{(\pi/2,\ \pi/2)}^{(\pi,\ \pi)} (\cos y + y \sin x)\, dx + (\cos x + x \sin y)\, dy$$

   Answer: 0

4) Find a potential function for the conservative vector field: $F(x, y, z) = (y + z)i + (x + z)j + (x + y)k$.

   Answer: $f(x, y, z) = xy + xz + yz$

5) Find the work done in moving a particle from (0, 0) to (2, 4) along the parabola $y = x^2$ against the force $F = x^2 y i + y j$.

   Answer: $\frac{72}{5}$

15.4 Green's Theorem
Solve the problem.

1) Use Green's theorem to evaluate

   $$\oint_C P\, dx + Q\, dy$$

   $P(x, y) = xy$, $Q(x, y) = e^x$; C is the curve that goes from (0, 0) to (2, 0) along the x-axis and then returns to (0, 0) along the parabola $y = 2x - x^2$.

   Answer: $\frac{8}{3}$

2) Use Green's theorem to evaluate

   $$\oint_C P\, dx + Q\, dy$$

   $P(x, y) = e^x - y^2$, $Q(x, y) = x^3 + \cos y$; C is the circle $x^2 + y^2 = 1$, oriented counterclockwise.

   Answer: $\frac{3}{4}\pi$

3) Apply the corollary to Green's theorem to find the area of the ellipse with equation: $(\frac{x}{a})^2 + (\frac{y}{b})^2 = 1$.

   [Suggestion: this ellipse has parameterization r(t) = ai cos t + bj sin t, $0 \le t \le 2\pi$.]

   Answer: $ab\pi$

4) Use Green's theorem to calculate the work

$$W = \oint_C \mathbf{F} \cdot \mathbf{T} \, ds$$

   done by the force $\mathbf{F} = x^2\mathbf{i} + y^2\mathbf{j}$; C is the part of the parabola $y = x^2$ over the interval [0, 1]. Find the work done in moving a particle along C from (1, 1) to (0, 0) against this force.

   Answer: $-\frac{2}{3}$

5) Use Green's theorem to calculate the work

$$W = \oint_C \mathbf{F} \cdot \mathbf{T} \, ds$$

   done in moving a particle along the parabola $y = x^2$ from (0, 0) to (1, 1) against the force $\mathbf{F}(x, y) = x^2\mathbf{i} + xy^2\mathbf{j}$.

   Answer: $\frac{1}{21}$

6) Use Green's theorem to calculate the work

$$W = \oint_C \mathbf{F} \cdot \mathbf{T} \, ds$$

   done in moving a particle along the semicircle $x^2 + y^2 = 4$, $y \ge 0$, from (-2, 0) to (2, 0) against the force $\mathbf{F}(x, y) = x^2\mathbf{i}, -y\mathbf{j}$.

   Answer: $\frac{16}{3}$

7) Use Green's theorem to calculate the work

$$W = \oint_C \mathbf{F} \cdot \mathbf{T} \, ds$$

   done when a particle is moved along the helix x = cos t, y = sin t, z = 2t from (1, 0, 0) to (1, 0, 4π) against the force $\mathbf{F}(x, y, z) = -y\mathbf{i} + x\mathbf{j} + z\mathbf{k}$.

   Answer: $2\pi + 8\pi^2$

8) Use Green's theorem to transform $\int_0^1 \int_0^x x^2 \, dy \, dx$ into a contour integral and then evaluate the integral directly.

   Answer: $\frac{1}{4}$

9) The path C is the unit circle $x^2 + y^2 = 1$, oriented counterclockwise. Use Green's theorem to evaluate
$$\oint_C y^2\, dx + x^2\, dy$$

Answer: 0

10) Given: the circular disk D with boundary C whose equation is $x^2 + y^2 = 1$. Use Green's theorem to evaluate
$$\iint_D x^2\, dA.$$

Answer: $\dfrac{\pi}{4}$

11) Let C be the closed curve $x^2 + y^2 = 1$ in the xy-plane, oriented counterclockwise. Use Green's theorem to evaluate
$$\oint_C (e^x - y^3)\,dx + (x^3 - 4\sin y)\,dy$$

Answer: $\dfrac{3\pi}{2}$

12) The region D in the xy-plane is bounded above by the graph of the parabola $y = 1 - x^2$ and below by the x-axis. Use Green's theorem to evaluate $\iint_D 2y\, dA$.

Answer: $\dfrac{16}{15}$

13) Use Green's theorem to evaluate the contour integral
$$\oint_C 2xy\, dx + (x^2 - y^2)\, dy$$

C is the path consisting of the straight line segment from (0, 0) to (1, 0), the straight line segment from (1, 0) to (1, 1), the straight line segment from (1, 1) to (0, 1), and the straight line segment from (0, 1) to (0, 0).

Answer: 0

14) Use Green's theorem to evaluate the contour integral
$$\oint_C (2^x + e^y)\, dx + (4x^2 - e^y)\, dy.$$

C is the path consisting of the straight line segment from (0, 0) to (1, 0), the straight line segment from (1, 0) to (1, 1), the straight line segment from (1, 1) to (0, 1), and the straight line segment from (0, 1) to (0, 0).

Answer: 5 − e

15) Use Green's theorem to transform the integral $\int_0^1 \int_0^x y \, dy \, dx$ into a line integral and then evaluate the line integral.

Answer: $\dfrac{1}{6}$

16) Use Green's theorem to evaluate the line integral

$$\oint_C (xy + z^2) \, ds$$

where C is the straight line segment from (0, 0, 0) to (1, 1, 1).

Answer: $\dfrac{2\sqrt{3}}{3}$

15.5 Surface Integrals
Solve the problem.

1) Evaluate the surface integral $\iint_S f(x, y, z) \, dS$, given $f(x, y, z) = 2y + 3z + 4$; S is the part of the plane $z = 2x + y$ that lies inside the cylinder $x^2 + y^2 = 4$.

Answer: $16\pi\sqrt{6}$

2) Find the moment of inertia $\iint_S (x^2 + y^2) \, dS$, of the given surface S with respect to the x-axis. Assume that S has constant density $\delta \equiv 1$. S is the part of the sphere $x^2 + y^2 + z^2 = 16$ that lies above the plane $z = 6$.

Answer: $\dfrac{4096\pi}{81} \approx 158.864$

3) Given: $\mathbf{F} = 2x\mathbf{i} + 2y\mathbf{j}$; S is the hemisphere $z = \sqrt{16 - x^2 - y^2}$, evaluate the surface integral $\iint_S \mathbf{F} \cdot \mathbf{n} \, dS$ where **n** is the upward-pointing unit normal vector to S.

Answer: $\dfrac{512\pi}{3} \approx 536.165$

4) Calculate the outward flux of the vector field $\mathbf{F} = x\mathbf{i} + y\mathbf{j} + z\mathbf{k}$ across the closed surface S, the boundary of the solid paraboloid bounded by the xy-plane and $z = 9 - x^2 - y^2$.

Answer: $\dfrac{243\pi}{2} \approx 381.704$

5) A uniform solid ball has radius 4 and its temperature u is proportional to the square of the distance from its center, with u = 64 at the boundary of the ball. If the heat conductivity of the ball is K = 3, find the rate of flow of heat across a concentric sphere of radius 5

Answer: $-12{,}000\pi$

15.6 The Divergence Theorem
Solve the problem.

1) Use the divergence theorem to evaluate $\iint_S \mathbf{F} \cdot \mathbf{n}\, dS$ where $\mathbf{n}$ is the outer unit normal vector to the surface S.

   $\mathbf{F} = 3x\mathbf{i} + 2y^2\mathbf{j} + 4z\mathbf{k}$; S is the surface of the plane $x + y + z = 6$

   Answer: 468

2) Use the divergence theorem to evaluate $\iint_S \mathbf{F} \cdot \mathbf{n}\, dS$ where $\mathbf{n}$ is the outer unit normal vector to the surface S.

   $\mathbf{F} = 2z\mathbf{i} + x\mathbf{j} + y^2\mathbf{k}$; S is the surface of the region bounded by $z = 4 - x^2 - y^2$ and the xy-plane

   Answer: 0

3) Use the divergence theorem to evaluate $\iint_S \mathbf{F} \cdot \mathbf{n}\, dS$ where $\mathbf{n}$ is the outer unit normal vector to the surface S.

   $\mathbf{F} = (3x + \cos z)\mathbf{i} + (2y - \sin x)\mathbf{j} + (4z + e^y)\mathbf{k}$; S is the surface of the region bounded by the planes $z = 0$, $y = 0$, $y = 4$, and the paraboloid $z = 2 - x^2$.

   Answer: $96\sqrt{2}$

4) Use the divergence theorem to evaluate $\iint_S z^2\, dS$ where S is the spherical surface $x^2 + y^2 + z^2 = 1$.

   [Suggestion: the problem here is to recognize z^2 as $\mathbf{F} \cdot \mathbf{n}$ for some vector function $\mathbf{F}$, where $\mathbf{n}$ denotes the outer unit normal vector to S.]

   Answer: $\iint_S z^2\, dS = \iint_S \mathbf{F} \cdot \mathbf{n}\, dS$ where $\mathbf{n} = x\mathbf{i} + y\mathbf{j} + z\mathbf{k}$ and $\mathbf{F} = z\mathbf{k}$. Then by the divergence theorem,

   $\iint_S z^2\, dS = \iiint_V \text{div } \mathbf{F}\, dV = \iiint_V 1\, dV = \frac{4}{3}\pi$

15.7 Stokes' Theorem
Solve the problem.

1) Use Stokes' theorem to evaluate $\iint_S (\nabla \times \mathbf{F}) \cdot \mathbf{n}\, dS$, where $\mathbf{F}(x, y, z) = 2y\mathbf{i} + 3x\mathbf{j} + e^z\mathbf{k}$; S is the part of the paraboloid $z = x^2 + y^2$ below the plane $z = 4$; $\mathbf{n}$ is the unit normal vector that point generally upward.

   Answer: 4π

2) Use Stokes' theorem to evaluate $\iint_S (\nabla \times \mathbf{F}) \cdot \mathbf{n}\, dS$, where $\mathbf{F}(x, y, z) = yz\mathbf{i} + xz\mathbf{j} + xy\mathbf{k}$; S is the part of the cylinder $x^2 + y^2 = 1$ that lies between the two planes $z = 1$ and $z = 3$; $\mathbf{n}$ is the unit normal vector that point generally upward.

   Answer: 0